METAL FINISHING TECHNIQUES

BY

ALEX WEISS

© 2009 ALEX WEISS

All rights reserved. No part of this publication may be reproduced, stored in a retrieval system, or transmitted in any form or by any means, electronic, mechanical, photocopying, recording or otherwise without prior permission in writing from the publishers.

Alex Weiss asserts the moral right as author of this work.

British Library Cataloguing-in-Publication-Data: a catalogue record of this book is held by the British Library.

First printing 2009

ISBN 978-0-9564073-1-3
CAMDEN MINIATURE STEAM SERVICES
Barrow Farm, Rode, Frome, Somerset, BA11 6PS
www.camdenmin.co.uk

Camden stock one one of the widest selections of engineering, technical and transportation books to be found; contact them at the above address for a copy of their latest free Booklist, or see their website.

Layout design by Camden Studios

Printed by Imprint Design & Print, Newtown, Powys

CONTENTS

	PAGE
ACKNOWLEDGMENTS	v
INTRODUCTION	**01**
Machine and hand finshes	05
Cleaning and pickling	07
Choosing the finish	08
Safety considerations	09
About this book	10
CHAPTER 1: GRINDING	**11**
Abrasive materials	12
Bonded abrasives	12
Grinding machines	16
CHAPTER 2: SANDING	**23**
Sanding abrasives	23
Coated products	24
Super abrasives	26
Bonded abrasives	27
Non-woven products	27
The sanding process	29
Sand blasting	30
CHAPTER 3: BUFFING AND POLISHING	**33**
Buffing machines	34
Buffing and polishing	37
Other polishing methods	40
CHAPTER 4: HONING AND LAPPING	**43**
Honing	43
Lapping	48
CHAPTER 5: REAMING	**55**
Reamer types	55
Using reamers	59

CHAPTER 6: BROACHING — 61
Types of broach — 62
Making a broach — 65
Using broaches — 65

CHAPTER 7: BURNISHING AND SCRAPING — 69
Burnishing — 69
Scraping — 73

CHAPTER 8: BARE METAL FINISHES — 77
Planishing and peening — 77
Chasing and repoussé — 78
Engraving — 79
Engine turning — 81
Brushed finish — 82
Knurling — 83
Oil or grease coating — 84
Other rust barriers — 85
Etching — 86

CHAPTER 9: METAL COLOURING — 87
Bluing and blackening — 87
Electroplating — 89
Electro-less plating — 90
Anodising — 90
Applying patinas — 91

CHAPTER 10: PAINTING — 93
Paint finishes — 93
The equipment needed — 94
Surface preparation — 98
Paint choices — 101
Specialist coatings — 105
Applying the paint — 106
Cleaning and storage — 113

CONCLUSIONS — 115

USEFUL CONTACTS — 117

INDEX — 119

METAL FINISHING TECHNIQUES   03

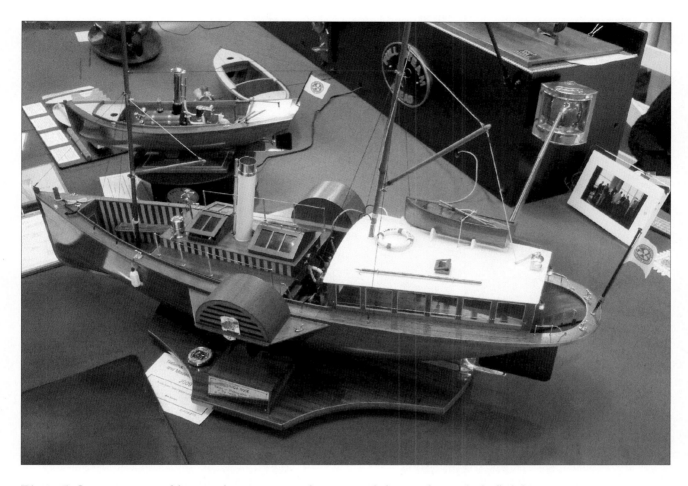

Photo 5: Steam-powered boats place extra environmental demands on their finishes

Photo 6: A marine steam engine with painted parts, polished brass parts and bare steel ones, as well as less visible bearings, crossheads, cylinders, pistons and valves that may need to be lapped or honed

04 ■ INTRODUCTION

Photo 7: Clocks are commonly built from steel and brass, with the latter polished and protected, and the steel pivots burnished

prototype. A particularly fine finish may be required for piston/cylinder interfaces, cross heads and bearing surfaces. Special action may be needed to ensure a correct fit by reaming, honing, lapping, burnishing, scraping or using a broach. Often a bare-metal finish is required and buffing and polishing the metal are popular options. Occasionally, there is a need to decorate the metal by further working, when chasing or repoussé, engraving, engine turning, etching, frosting or knurling may be appropriate. Sometimes electro-plating, anodising, or adding a patina may be suitable finishes. Finally, what is probably the most widespread type of finish – painting – requires a number of choices to be made. Which, from the huge range of paints, varnishes and lacquers, should be used and what is the most appropriate method of application.

Photo 8: The lathe turning marks are visible on the rim of the cast iron flywheel

Photo 9: The lighting exaggerates the finish of the milled surface of the top of this component

Regardless of which techniques are used, it is essential to have the best possible light when working on a project. Natural light is always preferable but is often inadequate. Quality artificial lighting is therefore essential to achieve the finest results; ideally fluorescent tubes for background lighting and adjustable spotlights for detailed work.

The measurements in this book are first given in metric form. They are followed in brackets by their Imperial equivalents in decimal form. Some of these equivalent figures are rounded, rather than absolutely exact.

MACHINE AND HAND FINISHES

Turning and milling are the traditional way of achieving both a fine and precise finish. The effective limit of such machining activities is a final cut of around 0.025mm (0.001") with a good surface appearance. Hand finishing of some awkwardly shaped items may occasionally prove essential but the finish so produced is rarely a match for a well-machined one.

TURNING
Metal carefully turned in the lathe produces a finish that is quite adequate for many purposes. But close surface examination shows its appearance may be less than perfect, the results depending on:

1. The metal being turned.
2. The speed of the chuck.
3. The rate of feed of the tool and the depth of cut.
4. The smoothness of the feed.
5. The cutting tool, its alignment, shape, sharpness, ability to stay sharp and its match for the material being cut.
6. A properly-maintained quality machine tool.
7. A good supply of cutting fluid.

The finish may or may not be acceptable from an aesthetic point of view but for a bearing surface it will almost certainly prove unsatisfactory. There are always surface imperfections and these can been seen in Photo 8. It is the final cut that determines the finish achieved. For the best results, a light cut with a sharp tool and a slow but consistently steady rate of feed along or across the work piece should be used, depending on whether the task is to turn or to face.

MILLING
Milling produces a different surface finish from turning that also depends on the type of material being worked, the kind of milling tool being employed, the state of its cutting edges, the selected velocity of the mill, an adequate supply of coolant, the amount of metal being removed with each pass of the tool and the speed and smoothness with which it is moved across the surface being milled. An example of the finish left after milling is illustrated in Photo 9. As with turning, it is the final cut that establishes the quality of the finish. A sharp tool, a light cut and a constant but steady feed rate are needed to obtain the optimum outcome.

Photo 10: Reflections are visible in the fine surface finish that results from flycutting on a vertical mill

FLY CUTTING

Fly cutters are rotating tools, used in milling machines or on lathes, to finish large flat surfaces. A key benefit of fly cutting is that the tool can cut the whole surface being machined in a single pass on all but the largest parts. Most fly cutters employ a simple single-point cutting tool, not dissimilar to a lathe tool, which is mounted in a special holder. Fly cutters can remove large amounts of metal and level it, with the advantage that the cutting bit is easily re-sharpened with a grindstone. Photo 10 shows an excellent finish can be achieved, though the original photograph does show the curved nature of the cuts.

DRILLING

Drilled holes rarely have a fine inside finish though for many purposes it is satisfactory. The internal quality can

Photo 11: A deburring tool with easily replaceable bits is useful for cleaning rough edges

Photo 12: Saw marks on a brass block, typical of those produced by a powered hacksaw

be improved by drilling the hole a little undersize, then increasing the diameter of the drill so the required size is achieved. The final two passes should be made with drills that differ by only 0.5mm (0.02") in diameter. All the same, for a precision hole with an accurate diameter, reaming the hole is essential to produce a better internal finish that will be suitable for use as a bearing.

DE-BURRING

Drilled holes, turned work pieces and milled items often end up with sharp edges that need de-burring. There are many ways of achieving this, such as filing, sanding or using a custom de-burring tool. It is important to remove any rough edges with some care to avoid any deterioration to the remainder of the existing finish.

SAWING

As Photo 12 shows, even when done with care, sawing through metal invariably leaves a rough edge that then requires further work to provide a smooth finish. Either machining or filing will be needed to flatten the surface and eliminate the ridges caused by the saw.

FILING

A well hand-filed finish, with draw filing to remove the tool marks, can produce a remarkably good finish. Rough work can progressively be smoothed with a coarse, then a medium and finally with a fine file.

Draw filing helps to remove marks created by the previous cross-filing pass. A fine file, held at right angles to the work, is pulled back and forth over the work surface, preferably using the handle end that is normally the sharpest and placing the first two fingers over the top

1: GRINDING

Grinding is a process where abrasive material is bonded to form a wheel that is rotated against a component to remove small amounts of metal. The formation of chips during grinding is similar to that in other machining processes. However, grinding generates lots of heat, indicated by the creation of quantities of sparks with ferrous metals, so that keeping the work piece cool is essential. Grinding is primarily used in the home workshop to sharpen hardened tools and industrially to produce precision components. Unfortunately grinding does not work well on soft metals.

The technique of grinding may be employed to remove precise but minute quantities of metal from components in any workshop if suitable equipment is available. The process can provide a quality finish with a tool-post or surface grinder, though the use of the latter in home workshops is still relatively rare However, at least one milling machine of Far-East manufacture provides a surface-grinding capability.

As a finishing process, grinding is used to improve surface appearance, abrade materials too hard to be machined and tighten tolerances (better than ±0.0025 mm [±0.0001"] for steel) on flat and cylindrical surfaces by removing very small amounts of material. This degree of accuracy is hardly a normal requirement for any model engineer!

In addition, using a tool-post grinder fitted to a lathe is an excellent way of applying grinding to improve the surface finish of circular components, particularly on items like crankshaft bearings.

There are several types of bench grinder powered by electric motors and suitable for the home workshop. They remove a lot of hard manual work from the process of sharpening tools though the increasing popularity of replaceable carbide tool tips is reducing the need for sharpening. But when the edge of any tool being ground is overheated, it burns and turns to a bluish colour as a result of excessive heat build-up in the tool. This causes the metal to lose its temper and its cutting edge. And once this has happened, the edge needs to be ground away to remove all the damaged metal before a new edge is sharpened.

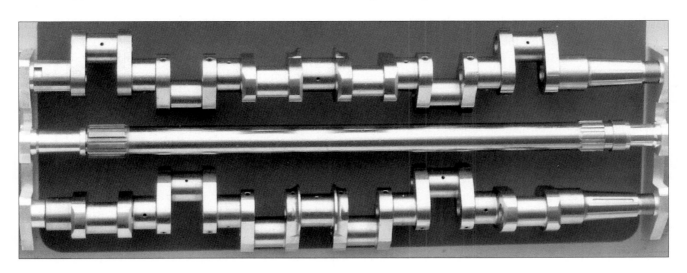

Photo 1.1: Two crankshafts for a model internal-combustion engine with beautifully ground bearing surfaces

Photo 1.2: An early limestone grinding wheel made around 2,000 BC

ABRASIVE MATERIALS

Originally, grindstones were hewed from solid blocks of sandstone, mounted on a shaft over a trough of water and turned by hand. The liquid removed worn particles of the stone that came free. However today, common substances used as abrasives include alumina or corundum, carborundum and less commonly emery. Aluminium oxide is found as a natural crystalline material called corundum that is readily converted into synthetic crystals; an abrasive called alumina. These days most corundum is manufactured from a uniform aluminium oxide. It is a tough material particularly in compression that resists wear, is excellent for working ferrous metals and normally comes in a range of grain sizes from 100 to 600 mesh.

The grain sizes used to make most wheels divide into four major categories from coarse to very fine. Separating screens sort the grains into different sizes. The size number corresponds to the number of meshes in the screen per linear inch (no metrication here yet). Coarse is 12 – 24, medium 30 – 60, fine 70 – 120 and very fine 150 – 240. However, the coarsest grinding wheels may have grain sizes as large as 8 mesh.

Emery is a mixture of corundum and magnetite or haematite that is found in the countries of the north-east Mediterranean. In the naturally occurring material, the grains are of irregular size giving a varying grinding performance unless graded by their dimensions. The range of sizes varies from the finest at 220 mesh to the coarsest at 20 mesh.

Carborundum is the widespread trade name for silicon carbide, the grains of which are harder and fracture less easily than those of alumina. It is a crystalline blue/black man-made material and is an excellent abrasive, widely used to make wheels for grinding. Grain sizes range from 60 to 3,000 mesh. Green grinding wheels (sometimes also black) are made from silicon carbide and are designed for sharpening tungsten-carbide tools.

Diamond and cubic boron nitride (CBN) are harder than aluminium oxide or silicon carbide and their cost is significantly higher, making their use in grindstones rather less common in home workshops. However, while diamond is the hardest known material, it is not a good abrasive for steel as it experiences a destructive chemical reaction during grinding; the same is true of silicon carbide.

All abrasives ultimately wear and their grains split into smaller bits. And as the grains are brittle, their cutting edges tend to fracture during use. The resulting smooth or jagged edges depend on the type and nature of abrasive product. Grains firmly fixed in place, as they are in a grindstone, will ultimately flatten. Any that are able to rotate, as they can during lapping, will become rounded.

These abrasive materials are also widely used for sanding, described in the next chapter and for honing and lapping covered in Chapter 4. There are even finer abrasives than those mentioned here that are used for sanding, buffing and polishing. These include several materials such as garnet, rouge, crocus and tripoli that are all described in Chapters 2 and 3.

BONDED ABRASIVES

Grinding has been used to sharpen tools since the earliest times of metal working. Certainly Bronze-Age man used sandstone wheels to sharpen swords, axes and the heads of spears and arrows. Photo 1.2 shows an Egyptian grindstone dating back some 4,000 years. And today, a bench grinder that operates on similar principles is one of the most-important powered tools likely to be purchased by any budding model engineer.

Grinding metal is a cutting process where the grains all cut the metal as they become exposed on the surface of the wheel. Modern grindstones are made by bonding selected and carefully sized grains of a chosen abrasive to form the shape of wheel required; normally cylindrical but differently shaped for specialised tasks. Grinding is capable of providing an exceptionally fine finish with a high degree of accuracy. A number of factors that impact on the ability of the stone to cut include:

METAL FINISHING TECHNIQUES ■ 13

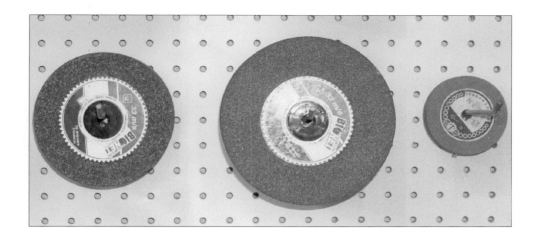

Photo 1.3: Three different-size grindstones for use on a variety of grinding machines

1. Type of grain material.
2. Grain size.
3. Individual shapes of the grains.
4. Grain orientation.
5. How easily grains crumble.
6. Type of bond employed.
7. Bond hardness and strength.
8. Percentage of grains to bond.
9. Cutting speed.
10. Depth of cut.
11. Feed rate.
12. Type and amount of coolant.
13. The way the wheel has been dressed.

It is a lot of variables and it makes the selection of the right wheel for any particular application very important. An indication of the wrong wheel for a specific grinding task is that it rapidly dulls and heats up the work piece too quickly.

GRAINS

As mentioned earlier, grain sizes vary from coarse, as low as 8 mesh to very fine up to 3,000 mesh or more.

BONDING

The grains of a wheel may be bonded together with ceramic or vitrified material, phenolic resin, rubber,

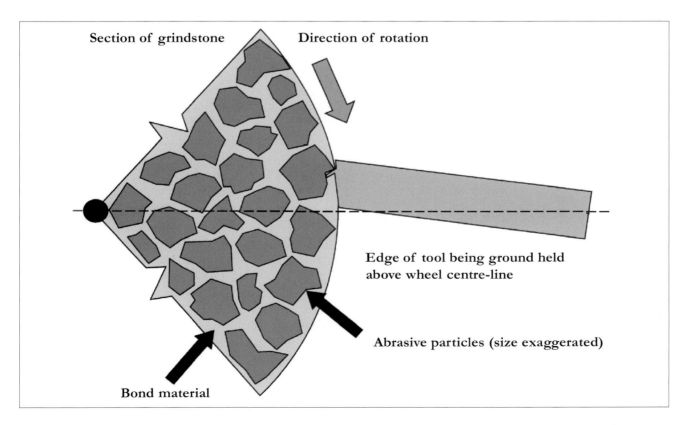

Drawing 1.4: The cutting process as metal, which must be held above the centre-line, is ground

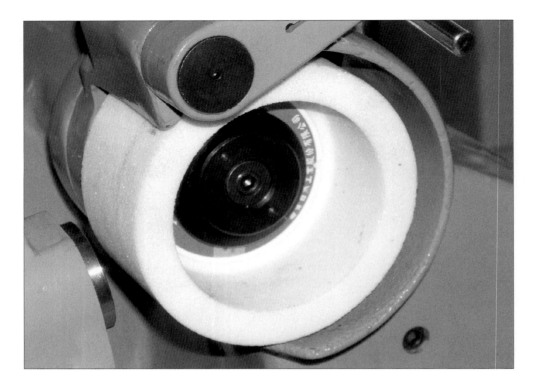

Photo 1.6: A classic U-shaped grindstone fitted to a tool and cutter grinder

shellac or even metal. Some bonding materials provide a rigid wheel, while others like rubber and shellac result in more flexible wheels that have a limited lifetime. Vitrified wheels are by far the most common for model-engineering use and are made by exposing a mix of the ingredients first to high-pressure moulding and then to firing at up to 1260°C. The abrasive can either be bonded with its particles close together or with significant space between the grains. The structure of a wheel is measured on a scale of 0 – 14. Zero indicates very close grains when the wheel is bonded, with 14 having the widest spaces between each of the grains.

The hardness or grade of a grindstone depends on its bond strength. A very-hard wheel holds the grains in place against the heaviest grinding forces. A soft wheel releases grains with just a light force. Thus the type and amount of bond determines a grindstone's grade; alphabetically in order of hardness, very soft: A – D, soft to medium: E – L, medium to hard: M – T, hard: U and very hard: V – Z.

The wheels supplied with most bench grinders have a hardness classification around N, considered hard. This class of wheel is very good for removing heavy nicks, such as those picked up by a cold chisel but can easily burn HSS tools. A very-hard green (occasionally black) silicon-carbide wheel is essential for sharpening carbide-tipped cutting tools.

The softer the wheel the less material is removed in a given time. The wheel slowly but continuously releases grain and thus presents new cutting edges. This prevents overheating and produces a better edge. New bench grinders usually come with a wheel on each end; one coarse and the other a fine hard-grey aluminium-oxide wheel. Two other useful wheels are a green silicon-carbide wheel and a soft pink aluminium-oxide one.

Diamond and cubic-boron-nitride (CBN) wheels are not made by bonding grains of the material to form a complete wheel because of the expense. Rather they are made by resin-bonding grains to a suitably shaped metal former. And the layer of diamond or CBN may only be 10mm wide by 3mm deep. Drawing 1.5 shows examples of typical wheel cross-sections.

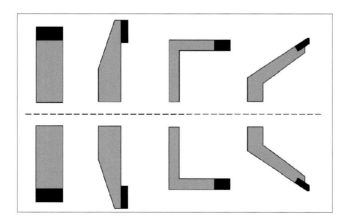

Drawing 1.5: Four useful grindstone cross-sections showing the diamond or CBN parts in black

Photo 1.7: The left-hand wheel has a speed limit of 4,460 rpm, the right-hand one, less conveniently marked, indicates a maximum peripheral speed of 33 metres per second (107 feet per second)

SHAPE AND SIZE

Wheels come in a range of sizes and shapes. The most popular are disc-shaped of 150mm (6") diameter but both larger and smaller diameters are readily available, depending on the capabilities of the grinding machine. Other shapes include flat discs, cones, taper and dish cups; the last two commonly used on tool and cutter grinders. All grinding wheels have a speed limit which depends on their diameter and construction. The design top speed is normally marked on the side of the stone. Exceeding it can cause the wheel to explode. And any wheel that shows signs of disintegrating at the circumference should be discarded for the same reason.

TRUING AND DRESSING WHEELS

Few brand-new wheels these days need truing, that is adjusting their circularity, due to quality control at the manufacturing stage. However, the shape of their cutting faces may need correcting. And after some use, any wheel will need the circumference of the wheel made concentric to its spindle and its face made flat. These tasks can be undertaken at the same time as dressing, removing any metal that is clogging the cutting surface, to reinvigorate the cutting action. Photo 1.8 shows an inexpensive rotating-wheel dresser that works well, but the diamond-stone dressers in Photo 1.9 are much more useful tools. They can be used to dress a wheel to a high degree of accuracy; essential when using a tool-post, surface or cylindrical grinder or a tool and cutter grinder. In all cases, the wheel dresser is held firmly at or just below the centre-line. Both rotating-wheel and multi-point diamond dressers should be held horizontal. However, a single-point diamond dresser needs to have some 10°

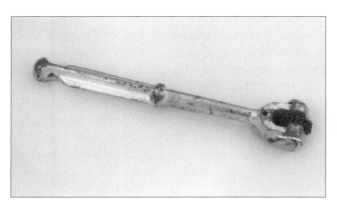

Photo 1.8: A low-cost dressing tool for grindstones

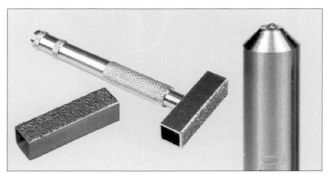

Photo 1.9: A multi-point diamond wheel dresser with spare head and, right, a single-point dresser. Photo courtesy Arc Euro Trade

CHAPTER 1: GRINDING

Photo 1.10: A classic 150mm (6") bench grinder fitted in a wooden case to help contain the dust. The custom clear-plastic guard is open for this photograph

of negative rake to avoid chatter and a side angle of some 30° as it is moved sideways across the wheel until clean and true. Unfortunately, the process produces a lot of abrasive dust that will coat other machine tools and models in the vicinity. They should therefore be covered before dressing any wheel in the workshop.

Photo 1.11: The Brierley TG10 bench grinder with stand and dust collector. Photo courtesy Chester Machine Tools

GRINDING MACHINES

Methods of sharpening tools range from hand-holding the work piece against a bench-grinding wheel to the use of a sophisticated tool and cutter grinder that will ensure the most appropriate shape when grinding lathe tools and will also correctly sharpen end and slot mills. Machine tools such as cylindrical and surface grinders will only rarely find their way into the home workshop, in part due to their size and weight, in part due to their single function and cost. Surface grinders may, however, be available to those undertaking evening classes, in which case the class instructor will be able to provide training in their use. However, tool-post grinders can be fitted to most lathes; any if home-made or appropriately adapted.

BENCH GRINDERS

It is hard to imagine any workshop equipped with machine tools that does not include a bench grinder. These are invariably double-ended machines that can be concurrently fitted with two different grindstones. A typical example is shown in Photo 1.10 that has been mounted in a dust-retaining box to minimise the amount of abrasive flying around the workshop. In any case, it is wise to mount any grinding machine as far away as possible from other machine tools.

The ubiquitous bench grinder, fitted with a relatively coarse wheel at one end and a fairly fine one at the other

2: SANDING

Sanding is commonly associated with wood working, using increasingly fine abrasives usually mounted on a flexible backing, such as cloth or paper, to obtain the required appearance. However, it is also a valuable metal-finishing technique and modern abrasives may be mounted on a range of materials such as nylon fibres or steel sheet as well as on paper or cloth.

Depending on the grade of abrasive and the skill of the worker, hand methods can produce a very fine finish. However, machinery such as linishers and disc sanders, often combined in a single machine, are relatively low cost and are increasingly finding their way into home workshops. There are several types of electric-powered sanders that remove the hard manual work from the task of metal smoothing. Abrasive products used for sanding metal can be categorised into several main groups:

1. Coated abrasives such as belts, discs and paper or cloth sheets.
2. Metal-backed super abrasives employing quality grains, such as diamond or cubic boron nitride (CBN), are laminated to a metal backing using a phenolic-resin bond. Alternatively tungsten-carbide grit is brazed to a steel backing.
3. Bonded abrasives comprise an abrading material contained within a matrix.
4. Non-woven abrasives come in the form of pads, discs, belts and wheels.

All four types of abrasive can be used to remove material during metal working. Grit sizes for coated abrasive are normally categorised in Europe into the following five grades:

Coarse	40-60
Medium course	80-100
Medium	120-150
Fine	180-220
Very fine	240 upwards

It is worth noting that automobile-finishing work employs grit sizes as fine as 2,500. All abrasives will eventually erode and their granules will divide into undersized pieces. As the grains are brittle, their cutting edges will tend to break during the sanding process. Whether these fractures leave smooth or sharp edges will vary with the nature of the abrasive and the type of abrasive product. And any granules fixed in orientation will eventually become flat.

Super abrasives tend to come in a wide range of grits; from 80 to around 750 grit and, with their solid metal backings, may be thought of as stones used for honing that are described in Chapter 4.

SANDING ABRASIVES

Common substances used for sanding include alumina or corundum in grain sizes from 100 to 600 mesh, emery

P Grade (coated)	Particle size range in microns
P240	58.5 ± 2.0
P280	52.2 ± 1.5
P320	46.2 ± 1.5
P360	40.5 ± 1.5
P400	35.0 ± 1.5
P500	30.2 ± 1.5
P600	25.8 ± 1.0
P800	21.8 ± 1.0
P1000	18.3 ± 1.0
P1200	15.3 ± 1.0
P1500	12.6 ± 1.0
P2000	10.3 ± 0.8
P2500	8.4 ± 0.5

Table 2.1: Abrasive grades above 220 and micron size equivalents. Note that 1 micron = 0.00004"

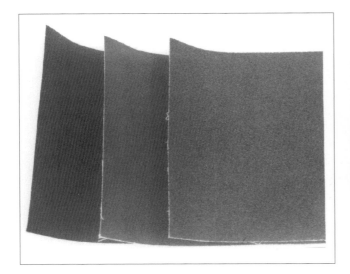

Photo 2.2: Three different grades of emery cloth that provide long-lasting qualities

Photo 2.3: Sanding belts come in a wide range of widths, lengths and grit sizes

that varies from 20 to 220 mesh while carborundum or silicon carbide and diamond have grain sizes from 60 to 3,000 mesh. These materials have already been described in Chapter 1. Garnet, employed on fine-coated products, is a naturally-occurring silicate that in a pure and flawless state may be cut as a gem stone. Of the many forms of garnet, iron and aluminium garnet are reddish brown and are best suited to achieving very fine finishes.

COATED PRODUCTS

For sanding, abrasives are mostly coated onto cloth or paper in the form of flat sheets, rolls, belts, or discs as well as in the form of flap wheels. Coated products should always be considered as consumables since their cutting power constantly and fairly rapidly declines throughout their life until they are discarded.

They are undoubtedly the most widely-used abrasive products for a myriad of tasks in the home workshop. The abrasive grains, normally of aluminium oxide, silicon carbide or garnet, are bonded to the backing using resin or glue. Grain sizes used for roughing range from 24 – 60, for intermediate work from 80 – 150 and for finishing from 180 – 1,200.

The backing may be paper that is graded in increasing thickness from A to F. For cloth the less-logical sequence of increasing thickness is J, X, Y, T, and M. Cloth made from cotton, polyester, rayon or Mylar provides better flexibility and longevity than paper. Coated products normally have the grit size and weight of backing printed on the back e.g. 60J.

FLAT SHEETS

Quite a few applications will need the use of paper or cloth abrasives. Cloth provides the greatest flexibility and durability. Cloth-backed or waterproof-paper-backed abrasives are needed for wet and dry requirements as ordinary paper-backed ones will rapidly disintegrate if wetted. Cloth and paper products tend to come in standard sheet sizes of 280 x 230mm (11" x 9"), though smaller sheets are also popular. And for larger pieces, rolls of sheet material may be purchased. Sanding blocks are also widely available with a range of different backings for the coated material. Sheets may be obtained with grits ranging from 24 – 1,000 with the grit size marked on the reverse side. Pre-cut discs (and other custom shapes) are also popular for use with disc-sanding machines, often with a self-adhesive backing.

BELT AND DISC SANDERS

Belt sanders or linishers employ a continuously turning belt running over a flat metal-backing plate which may be vertical, horizontal or at an angle between the two. They are often also fitted with a disc sander (or a grindstone) on the other end. The disc may be fixed to the circular metal-backing plate with an adhesive, or Velcro to simplify replacement, but it is essential that the sanding disc is installed absolutely flat. Disc sanders can also be mounted on pads and fitted to electric drills. Orbital sanders that use an oscillating abrasive sheet and drum sanders are designed for wood-working as are so-called power files that can also find uses removing metal. Belts come in varying widths to fit the sander and with different abrasive grades while discs vary in diameter to suit the machine.

Depending on the grade of abrasive on the belt/disc, any powered sanding machine can remove metal remarkably

METAL FINISHING TECHNIQUES • 25

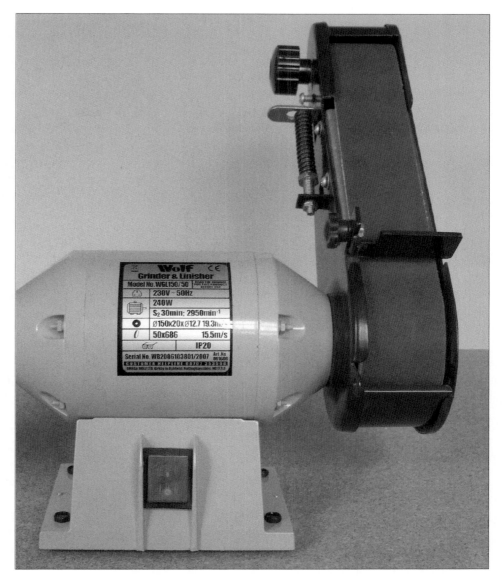

Photo 2.4: A belt linisher is an brilliant tool for removing and finishing metal

rapidly or, with a fine abrasive can provide a first-class finish. Nevertheless, hand working with emery cloth is still often the last stage in getting the parts of any model ready for the final finishing process. However, care is needed to avoid accidentally removing too much metal with a power sander in places where it is still needed.

FLAP WHEELS

Flap wheels are designed for rapid metal removal and are readily mounted in a drill chuck. They are ideal when

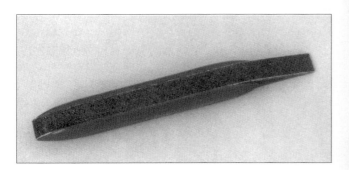

Figure 2.5. A sanding stick that employs a small belt of emery paper

Figure 2.6. A typical low-cost disc sander

Photo 2.7: A flap wheel of coated paper showing its size and maximum-speed rating

dealing with small contact areas on components and an advantage is that their cutting power does not reduce as they wear down. They are perfect for blending, finishing and cleaning inside and outside diameters of all metals; de-burring and removing flash and parting lines from castings. Flap wheels are ideal for preparing work for buffing and polishing, particularly on complicated and awkwardly shaped items.

The flaps are made from abrasive strips radiating from a centre hub and are produced in many different sizes and shapes. Diameters start at a few tens of millimetres (1") and can range up to over a metre (39"). Flap wheels have an arbor-mounted hub that may be made of metal or cardboard and which uses an adhesive to hold the flaps in place. The usual range of abrasives shown on Page 12 may all be employed to make flap wheels.

Photo 2.9: A diamond sharpener that is really a hone

SUPER ABRASIVES

A somewhat different form of long-life abrasive is made by using diamond, cubic boron nitride (CBN) or tungsten carbide grains attached to a steel backing. In the case of diamond and CBN, they are permanently bonded to the metal backing with a phenolic-resin. Diamond-based products are excellent for providing a final finishing touch to carbide-tipped lathe tools but are more like hones than sanding tools. They come in a wide range of different formats as well as grit sizes and are described in Chapter 4.

Photo 2.8: Some of the large range of Perma-Grit sanding tools

METAL FINISHING TECHNIQUES ■ 27

Photo 2.10: Three different grades of tungsten-carbide abrasives on a steel backing

Photo 2.11: Garryflex blocks use silicon-carbide abrasive embedded in rubber

The flexibility of the block allows it to conform to curved or flat surfaces. A series of blocks is available in steps of abrasive coarseness from very coarse 36 grit to fine 240 grit. They may be used with or without a lubricant such as water or paraffin for a range of tasks including removing light rust or tarnish and providing a satin finish.

Tungsten carbide can be brazed onto steel and these products will cut both hard and soft metals. Unlike files, where the cutting edges become worn until the point when the files must be discarded, they have thousands of individual hard cutting edges giving them a longer life. They can cut in all directions, with wider grit spacing than emery cloth to help keep the cutting surface from clogging. They are available in different grades; ultra fine is about 420 grit, fine is around 320 grit, medium is approximately 220 grit, coarse is roughly 180 grit, extra coarse is around 120 grit and extra, extra coarse is approximately 80 grit. Perhaps the products most well-known to model engineers come from the range of super abrasives made by Perma-Grit and illustrated in Photo 2.8. Compared to conventional coated products, they are much longer lasting and will easily cut hard metals.

BONDED ABRASIVES

It is unusual to find bonded-abrasive products used for sanding, as opposed to grinding. A good example is Garryflex blocks that are made of a resilient rubber filled with a silicon-carbide abrasive grit that is evenly spread throughout the block. Thus performance does not deteriorate during the life of the block, unlike abrasive cloths and papers where the grit becomes finer with use.

NON-WOVEN PRODUCTS

Non-woven abrasives are made from a lattice of resin-bonded nylon fibres, impregnated with non-aggressive abrasive grains, to provide a cushioned, three-dimensional material that is flexible and can conform to the work piece. They are also both waterproof and washable. Non-woven abrasives come in the form of pads, discs, belts and wheels and may be used by hand or fitted to a suitable machine. These abrasives are primarily suited to surface-conditioning and light deburring tasks. They can be used for cleaning and removing corrosion as well as

Photo 2.12: Many different non-woven products are on sale at model engineering exhibitions

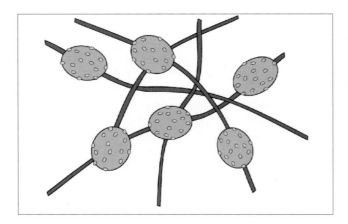

Drawing 2.13 Non-woven products are made from abrasive grains bonded to nylon fibres

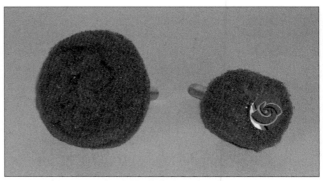

Photo 2.14: Two different shapes of non-woven shank-mounted sanding tools

providing the required appearance. The relatively gentle action of nylon and the abrasive grit used in non-woven abrasives make them great tools for light work. The abrasive grains are usually evenly dispersed throughout the nylon web, providing a continuous supply of new grains as the old ones and the fibres wear.

The density of non-woven abrasives refers to the amount of bonding agent and quantity of abrasive grains they contain. Harder densities usually cut faster, last longer and produce finer finishes than softer ones but softer ones conform better to the work surface with a lower tendency to load or burn the work piece. The size of the nylon fibres and the bonding agent used also have an impact. Fibre size significantly affects the cutting characteristics while the resin used to bond the fibres and anchor the abrasive can also alter them.

PRODUCT CATEGORIES

Non-woven products are defined by:

1. Their density that ranges from 2, open and most conformable, to 9 densest and most durable.
2. The abrasive used; 'A' or 'A/O' for aluminium oxide and 'S' or 'SC' for silicon carbide.
3. The grain size which falls into six broad categories. These are coarse (C) 50 – 80, medium (M) 100 – 150, fine (F) 180 – 220, very fine (VF) 240 – 360, ultra fine (UF) 600 – 800 and micro fine (SF) 1000 – 1200.

Thus a product might be listed as 5 SC VF – medium density with very-fine silicon-carbide grains.

TYPES AND USES

Non-woven abrasive products can be purchased in several different forms. Pads provide admirable flexibility and conformability when hand-finishing metal components.

Grit sizes range from coarse, used for general cleaning, to micro-fine for light scuffing and blending. In wet applications, a non-woven hand pad can be used as a non-rusting replacement for steel wool.

Many different non-woven discs can be employed for cleaning and finishing. Some discs are designed to operate on bench grinders. Coarse-grit discs can be used to remove surface rust, corrosion or light weld splatter. Right-angle discs may be fitted to portable grinders for cleaning, de-burring, blending and finishing, as well as for the removal of light rust, oxidation and coatings.

Non-woven belts are a useful substitute for conventional abrasive belts on bench machines when cleaning, buffing or polishing. The fact that these belts conform to the work piece minimizes any gouging.

There are three totally different types of wheel made from of non-woven materials:

1. Flap wheels employ sheets of non-woven material that are fastened radially around a central hub. These are ideal where conforming to a surface or obtaining a long-line brushed finish are important.
2. Convolute wheels use non-woven material wrapped around a centre core. These wheels can easily be shaped to match the outline of parts.
3. Unified wheels employ compressed multiple layers of non-woven web material bonded together to form a wheel. They are designed for general cleaning and de-burring.

Flap wheels and unified wheels may be run in either direction but convolute wheels must always be run in the direction indicated by the arrow marked on their side. The speed at which the wheel is run affects the finish, rate of cut, and wheel life. The faster the speed of the wheel, the harder the action and the finer the finish that

TASK	SURFACE SPEED PER MINUTE
Clean and improve surface	2,000 metres (6,500 feet) or slightly more
De-burr or remove oxide/rust	1,000 – 2,000 metres (3,250 – 6,500 feet)
Harsh buff metal surfaces	500 – 1,000 metres (1,625 – 3,250 feet)
Provide decorative finish	150 – 900 metres (500 – 3,000 feet)

Table 2.15: Recommended speeds for various sanding tasks using non-woven wheels

can be achieved. Slower speeds give a softer action and a coarser finish. Recommended speeds for common tasks are indicated in Table 2.15.

For most operations, use light to medium pressure, bearing in mind that flap wheels need much lighter pressure than other non-woven wheels. Unified wheels can survive the much higher pressures required for de-burring. But always avoid excess pressure that can deform the wheel and damage the work surface.

The grain size of an abrasive cannot alone be used to distinguish between the type of finish a non-woven wheel will produce. However grades of 80 and coarser tend to be considered suitable for relatively rapid metal removal, whereas those that are 120 and finer are associated with providing a comparatively fine finish. Both rapid metal removal and fine finishing may use a succession of stages with increasingly fine abrasive grain sizes to obtain the required final effect. In both of these processes, other factors including pressure, contact time and whether the process is undertaken dry or wet all affect the finish.

WIRE WOOL

Though not a non-woven product, steel wool has similar capabilities to these products, particularly for working softer materials. It comes in increasingly coarse grades: 0000, 000, 00, 0, 1, 2, 3, 4 and 5. The finer grades can be used for dulling gloss finishes and general cleaning, particularly of copper components prior to soldering; the coarser ones for tasks like rust and paint removal.

THE SANDING PROCESS

There are several decisions to be made before starting sanding.

1. Is the work going to be done by hand, by machine or a mix of both?
2. What type of sanding medium should be used? This will depend on the type of work to be undertaken and the shape of the component to be sanded.
3. Is a coated, bonded or a non-woven solution better for the selected application? (See Drawing 2.16.)
4. If a coated-abrasive, what backing material and what form: flap wheel, belt, disc, sheet or hand-held block or stick?
5. If a non-woven product, what shape; disc, flap wheel, other form of wheel or hand-held pad? Also what type of abrasive-holding material?

To start work, the component must be securely held, whether in one hand or a vice. Flat surfaces appear to be relatively easy to deal with; complex shapes often seem very difficult. It is probably hardest to sand a flat surface by hand as this requires both a firm backing to the sanding abrasive and a figure-of-eight sanding motion to avoid getting curved edges. Round components can be sanded with a strip of abrasive cloth held in both hands around the component and pulled back and forth. An abrasive that conforms to the shape of the components is often a good solution, particularly for components with complex

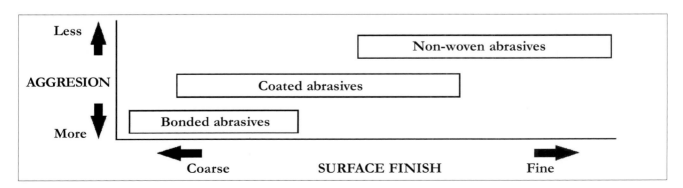

Drawing 2.16: The aggressiveness of various types of sanding products

Photo 2.17: The governor parts, except the balls, spring and vertical shaft, have all been sandblasted to give a matt-grey finish

profiles. Machine sanding is in many ways easier but care must be taken to avoid overheating and burning the metal or removing too much material as the process can be remarkably rapid.

The choice of the correct grade of grit is important and this will depend on the component's surface smoothness, or lack of it, and the final finish required. For components with a poor finish that needs improving, it is best to start work with a coarse abrasive grit. Then continue through several stages with intermediate grades, ending with a fine one for the best results.

Finally a choice of backing for a coated abrasive must be made; paper, cloth, plastic or steel in order of durability; also whether to use a lubricant; oil, paraffin or water with or without soap. In general, the use of a 'wet' approach will provide a finer result than the same abrasive used dry.

Employing mineral oil, paraffin or water as a sanding lubricant can improve the end results and this is easier with hand sanding or a using a linear machine than with a rotating one that can easily coat the workshop in liquid. The lubricant helps cool the abrasive and the work piece, assists removal of metal particles and avoids the formation of fine metal dust. Do not let the surface dry out and if excess wet particles of metal build up, remove them from the component with a dry cloth.

Always let the abrasive do the work avoiding any undue pressure that will only clog the abrasive and cause it to wear unnecessarily fast. And power sanding needs very little pressure.

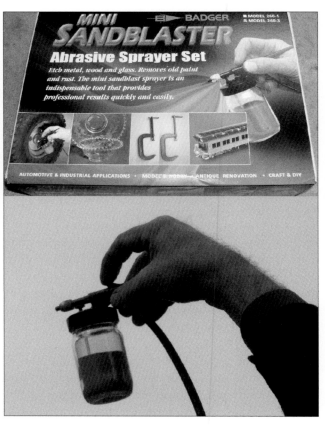

Photo 2.18: A mini sand blaster that is perfect for dealing with small components

Once the finest possible finish has been achieved with sanding, further improvements are possible by polishing and buffing that are described in the next chapter.

SAND BLASTING

Sand blasting as a method of cleaning, and particularly of removing paint, is a task readily carried out in the home workshop by anyone who has suitable facilities. The simplest solution for home-workshop use is a mini sandblaster that looks remarkably like an airbrush but sprays sand rather than paint. The fine sand impinges at high speed against the surface to be cleaned. The velocity is provided by compressed air directed through a nozzle. A small sand blaster is shown in Photo 2.18, and should be connected to an air compressor (see Page 96).

Sand blasting can be used to clean metal to a dull matt finish, even to the extent of removing rust and old paint. In addition to cleaning, the process can improve the finish and surface properties of the metal, smoothing and shaping it. At the same time the surface area of the metal tends to increase. However, some care is needed as the process can warp thin sections. Micro balls are a plastic equivalent to sand and are less aggressive on the metal but just as good at the cleaning process.

Sand blasting can provide a fine matt effect on metal; particularly on castings. And the bare metal finish can look particularly appropriate on some unpainted castings. Furthermore, it is often the ideal preparation prior to electro-plating, anodising or painting, which are described in Chapters 9 and 10. The use of abrasives such as aluminium oxide, garnet or silicon carbide for blasting are popular in industrial processes due to their longer life, albeit at the expense of increased nozzle wear.

A small booth can be made from a cardboard box, similar to the one for spray painting that is described on Page 97. It will catch the majority of the sand or micro balls for subsequent re-use. Take care not to use the sand blaster close to any machine tools in order to avoid getting abrasive sand particles onto the slides and other moving parts.

NOTES

modern technology to improve their performance. They can help to provide a fine polish on metal parts as can some electro-chemical methods that produce a shiny finish.

CUSHIONED ABRASIVES

Micro-Mesh is a flexible long-lasting latex-cushioned abrasive cloth. It has silicon carbide, aluminium oxide or diamond crystals as the abrasive, depending on the final use, attached with a flexible adhesive that allows the crystals to move and rotate without coming loose. It may be used on metal and will conform to the shape of the component being worked. Designed for polishing, when pressure is applied to the part being treated, the crystals slightly rotate to present an even array of sharp cutting edges with positive rake across the surface of the work piece. This provides a consistent scratch pattern and a high level of gloss or mirror finish; down to 0.0025mm (0.0001") or less. Excess pressure causes the crystals to descend into the cushioning layer, thus avoiding the production of scratches. Micro-Mesh may be purchased in the form of sheets, rolls, discs, belts or pads. There are four different types of Micro-Mesh: Regular (not designed for use on metal but good for most plastics), MX, Aluminium oxide and MXD.

Micro-Mesh MX is ideal for many metals and employs silicon carbide as its abrasive, apart from the two finest grades that use aluminium oxide. It is available in thirteen grades from 60 MX (around 300 grit) the coarsest, to 1200 MX the finest. Use wet it with a cutting or honing oil.

Micro-Mesh Aluminium oxide is perfect for polishing aluminium and may be used wet or dry. It comes in twelve grades from the coarsest 300 AO (equivalent to 150 grit) to the finest 12000 AO. Use the coarsest grade for scratch removal, 600 AO for a matt finish, 1500 AO for a satin finish, 4000 AO for a gloss finish and 12000 AO for a high-gloss finish.

Micro-Mesh MXD is suitable for any hard metal alloy and uses diamond as an abrasive in an ultra-flexible resin bond. There are fourteen grades that range from the coarsest 40 MXD to the finest 1800 MXD. 120 MXD is the equivalent to 180 grit.

MICRON-GRADED ABRASIVES

For the best results, micron-graded abrasives provide a finer finish than normal ones. 3M is one of several companies that manufacture micron-graded finishing films that are ideal for providing a fine polish. It is unfortunate that grain sizes, as detailed on Page 23,

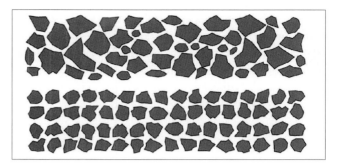

Drawing 3.17: Top, abrasive grains and bottom, micron-graded abrasive grains

decrease as the mesh number increases; the larger the number the smaller the grain size. However with micron grading, the figure refers to the size of the grains in microns (millionths of a metre, 0.00004") so that the smaller the number the smaller the grain size. And micron-graded abrasives are available in sizes down to less than a micron. It is perhaps regrettable that some of the finer finishing films are referred to as lapping films. As Chapter 4 explains, lapping involves the use of abrasive grains that are free to move.

ELECTRO-POLISHING

By suspending a component in a suitable chemical solution and connecting it to the anode of an electrical circuit, with a similarly sized cathode, the component will supply metal to the solution in what is a reverse-plating process. Electro-polishing is thus another way of polishing metals such as aluminium, brass, bronze, copper, nickel, steel and stainless steel. This electro-chemical method will remove high points from the metal surface to improve both its smoothness and sheen.

A well-cleaned part has to be immersed in a tank of suitable chemicals and connected to a high-output, direct-current power supply, quickly producing a well-polished part. Lead-tin-antimony cathodes and a power supply or battery are needed.

However, electro-polishing has serious limitations. First, it requires a slightly heated mixture of phosphoric and sulphuric acids; not recommended for home workshop use. The process generates both hydrogen and oxygen; a potentially explosive mixture. The component has to be cleaned in a pickle before polishing and then rinsed in de-ionised water afterwards. Silver is dissolved by the chemicals and thus silver-soldered components cannot be electro-polished. The process will not give a quality finish to a rusty or pitted component. And castings that tend to have dirt trapped in their pores cannot successfully be electro-polished. Any part must be sanded to a smooth finish if its surface is not in good condition and it must be rust, corrosion and dirt-free. 300 series stainless steels can be electro-polished to a mirror finish though other steels will still polish well. Aluminium shows best results if 99% pure or an alloy containing magnesium or magnesium and silicon. It is thus better to get an experienced professional company to carry out electro-polishing on components that need it.

4: HONING AND LAPPING

Although honing was historically associated with lapping (see Page 48) because both are used to polish away only a few hundredths of a millimetre (the odd thousandth of an inch) of material, today the difference between the two process is, by and large, well recognised. The main distinction between the two processes is that when honing, fixed abrasive in the form of a stone or stones are used to provide the finish. Lapping, on the other hand, involves an abrasive powder or paste that is used to finish a surface using a lap with abrasive spread on it, or to improve the way two components fit together.

It is common practice to provide the ultimate finish by honing metal surfaces that will move in physical contact with each other as well as to give the finest edge on cutting tools. Where pairs of components need to be a perfect fit and also have a superb finish, lapping of these items should follow after honing has been completed. This is described later in this chapter.

HONING

Honing is a method of obtaining a high degree of surface accuracy by the action of a fine abrasive stone or stones that retain the abrasive particles firmly in place. The stones are employed to remove minute amounts of material in order to smooth any irregularities in the metal surface. Since the cutting points of the honing abrasive grains are so tiny and large numbers of them can cut simultaneously, it is rare for significant amounts of heat or stress to be generated in the work piece. Thus honing minimises surface damage and provides a quality finish.

Honing can be carried out on most metals whether bar stock, castings or sintered items. Once drilled, reamed, bored, turned or milled to a precise size, the finish will depend on the abrasiveness of the hone, the metal being worked and the skill and care of the person undertaking the work.

Model engineers are most likely to come across honing as a method of improving the internal finish of the cylinder bore of steam or internal-combustion engines, as a way of providing a quality finish to plain bearings or as a method of sharpening tools. Honing may be carried out with specialist honing tools, oilstones, slip stones or even diamond hones. A set of cylindrical diamond stones for honing a cylinder typically lasts fifty times as long as other conventional stones and can produce a rounder, straighter hole. But not many model engineers will find that it justifies the extra expense; as much as twenty times the price! And diamond hones bring their own problems as well as advantages. Diamond has a tendency to plough through a metal surface rather than cutting through it. This will generate heat and distortion if the wrong pressure settings or lubrication are used.

Hones work well on metals such as bronze, brass, gunmetal and cast iron as well as mild or hardened steel and stainless steel. Lubrication of the honing stones is very important to avoid them picking up small fragments of the metal being honed. The lubricant will hold any particles in suspension as opposed to getting themselves embedded in the abrasive surface of the hone. Any metal driven into a honing stone may weld to the work piece, which will cause scratches or tears in the honed surface. Furthermore, in the worst cases, the honing stone may even chip thus seriously shortening or ending its life.

Many companies offer specially formulated honing oil but otherwise any light machine oil is an acceptable alternative. After or even better at regular intervals during use, stones must be wiped completely clean of the lubricant and any solid contaminants. Fresh oil should then be applied.

Vertical and horizontal honing machines are both used in industry and are able to provide the finest tolerances matched with superb surface quality on large volumes

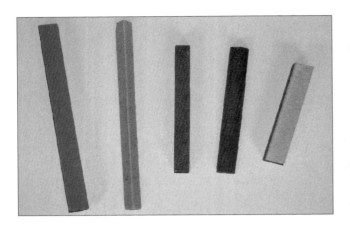

Photo 4.1: Slip stones come in a variety of shapes and sizes as well as degrees of abrasiveness

Photo 4.2: The three diamond hones in the centre are of different grades, while the two outer hones are small portable ones

of components. Such very specialist machines are most unlikely to find their way into any home workshops.

There is a technique that is almost a cross between sanding and honing. This is when a sheet of fine emery paper or cloth is placed flat on a smooth plate. Its surface may then be used to 'hone' the metal. The finish achieved depends on the grit size of emery employed, the quality of the flat plate, the consistency of the thickness of the paper and its emery layer.

FLAT HONES

Since the purpose of honing is to provide a fine finish, the stones employed are generally very fine ones with grits ranging up to 1,200. They are generally offered in one of three grades; coarse, medium and fine. Coarse stones are employed for sharpening very dull or nicked tools and remove metal quite fast. Medium stones are ideal for general work while fine stones should be selected where a very fine, keen-cutting edge is needed.

Oil stones are normally manufactured from aluminium oxide or silicon carbide, come in a range of different fine grits, and several brands are available with both coarse and fine faces. There is a full description of the various different abrasives on Page 12.

Stones may be sold as Arkansas or India stones, giving an indication of their origin. India oilstones are fine-grain blocks of dark-grey emery permeated with oil prior to use for sharpening metal tools. In order to get a keen edge on the tools, use oil to lubricate the rubbing surface and to prevent metal grains becoming driven into the surface of the stone.

Arkansas oilstones are silica rock, very hard with a fine grain and are a bluish or opaque white colour, while artificial oilstones are made from aluminium oxide. Small slip stones come in a variety of shapes; square, round, half-round, rectangular and triangular and should be employed to hone the cutting edge of lathe tools.

Diamond hones, where grains of industrial-grade diamond are bonded to a steel substrate, will hone any metal and most carbides and ceramics. They are quite affordable, are able to remove metal at a rapid rate and are long lasting in this application. These hones come in a range of sizes and often have a series of holes cut in them that catch the swarf that is formed. Ranging from super fine 1,200 grit through fine 600 and medium 400 to coarse 300 grit, these hones will provide a razor-sharp finish to tools and the smaller ones with handles can be used to hone awkward items like end and slot mills. It does not help that they are often referred to by retailers as diamond laps that 'eliminate the need for the continual application of loose abrasive'. Their main advantage is that they remain flat and do not require the use of any lubricant. However, when a significant amount of metal is being removed, it may be advantageous to use some lubrication such as paraffin to help clear away the chips of metal. Diamond hones may occasionally be washed with soap and water. In either case, it is important that they are carefully dried after use.

Other types of stones, mainly designed to sharpen metal wood-working tools include whetstones that are similar to oilstones but use water as a lubricant instead of oil, and water stones that, as their name suggests, are also lubricated with water but never with oil. They are readily available for honing metal with grits as fine as 8,000. Water stones are excellent at producing a very fine flat

METAL FINISHING TECHNIQUES

Photo 4.11: A close-up of two diamonds being lapped on a modern cast-iron lapping plate

LAPS AND LAP MATERIALS

For lapping to produce a plane surface, either a rotating table or a flat surface-lapping plate may be employed; the former is the norm for mass production but it is rare to find one in a home workshop. The latter is used for hand lapping and the plate should have slots or grooves machined in its surface to speed up the process and to help distribute the lapping compound. A piece of plate glass or mirror, placed flat on a workbench to prevent flexing, is a low-cost alternative.

When dealing with round parts, It is usual to work with an expanding lap and one that is suitable for external lapping is illustrated in Drawing 4.13. External laps

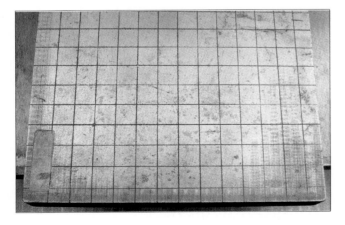

Photo 4.12: A well-worn cast-iron lapping plate that is in need of refurbishment

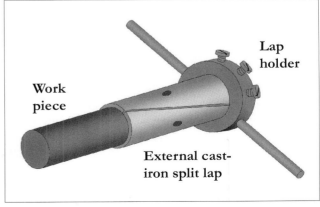

Drawing 4.13: A lap for working the outside of cylindrical objects

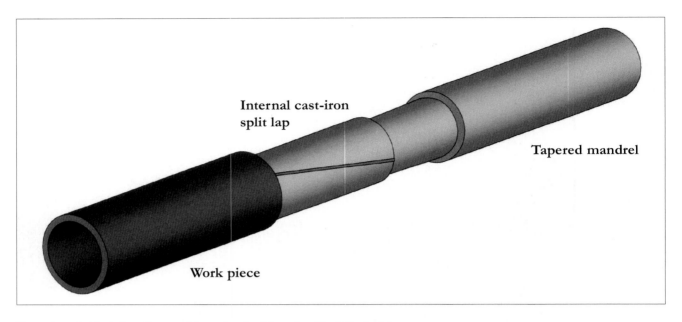

Drawing 4.14: A lap for working the inside of cylindrical objects

made with straight or spiral expansion slots require an external lap holder that works in a similar way to a die holder, with screws preventing the lap from rotating and allowing minute alterations to be made to its diameter.

Internal laps, also with straight or spiral expansion slots, should fit on a tapered arbor to expand the cylinder-shaped lap. The inside of the lap and the outside of the arbor have matching tapers. The exterior surface of the lap is straight and split by a narrow slot, allowing the lap to expand as it is moved up the arbor's taper. To make an internal lap requires the ability to turn accurate internal and external tapers and to split the lap using a slitting saw. An internal lap is shown in Drawing 4.14.

Unfortunately, a new internal or external lap is needed for each individual diameter that requires lapping; the amount of expansion available being extremely small.

The metal from which the lap is made must be carefully chosen to suit the material being lapped. The majority of model-engineering applications require steel to be lapped. In this case cast iron is the ideal material for the lap. As a result, cast iron is the commonest material for internal and external diameter laps as well as flat ones. Its composition enables abrasive to embed in its softer parts and to be held in place by the harder areas as the lap abrades the work piece. This is how a soft lap will cut harder items while minimising lap wear. Cast iron keeps its shape well and can produce an excellent finish.

Brass and copper laps are suitable for use on cast iron while copper is good for lapping brass or bronze. Brass lacks a structure that is able to retain the embedded abrasive in place, allowing the abrasive particles to be torn out as fast as they are embedded. This gives a slow cutting action. However, brass keeps its shape well and in some cases produces a fine finish. Although copper is considerably softer than brass or cast iron it maintains its shape. Its malleability makes it ideal for lapping soft materials without transferring abrasive into the work piece, resulting in the ability to produce first-rate results.

Lead was favoured in the past but is rarely used today. It does not hold the abrasive well and, in addition, lead grains are poison.

Some varieties of plastic such as nylon, Bakelite and filled plastics or even wood are occasionally used as laps. They can give an excellent finish. However, they all lack the ability to provide good geometry; whether circular or flat.

Diamond-coated laps are really hones, as they avoid the need to apply particles of abrasive. Diamond hones have already been described on Page 44.

Lapping should be used to take off some 0.025 - 0.05mm (0.001"- 0.002") of metal. It is essential that no abrasive is absorbed into the item being lapped and any traces are cleaned off with paraffin once lapping is completed. This is particularly true if a micrometer is employed to measure the final diameter as any grit can easily damage its anvil.

LAPPING COMPOUNDS

The quality of the lapped finish that is possible depends on the abrasive compound employed, its grit size and to

METAL FINISHING TECHNIQUES | 51

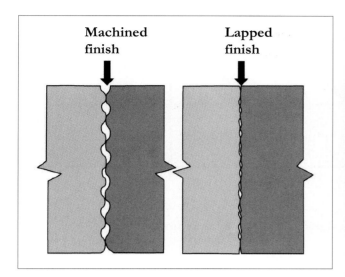

Drawing 4.15: A much-magnified comparison of a machined and a lapped finish

an extent the material being lapped. There is a large number of different lapping compounds made from diamond through silicon carbide to emery and crocus. Even liquid metal polish can be employed for the final rub. Diamond is the most expensive but gives the finest finish. It is essential to match the lapping compound to the metal that is to be lapped. Harder metals like cast iron and steel will need more aggressive abrasives than softer ones such as aluminium and bronze. Aluminium oxide is ideal for use on softer metals because of its grain structure, relatively high degree of hardness and sharpness. It will also cut hardened steels.

Silicon carbide does not fit the vast majority of model-engineering needs. It is only suited to lapping metals that have been hardened because it is extremely tough and sharp. However due to its brittleness it tends to break into very sharp pieces and these will produce scratches.

Using diamond lapping powder works well on any metal that is difficult to cut. Diamond powder is produced to fine micron sizes and can impart the finest finishes but at a price.

Lapping powder made of any of these abrasives may be spread directly to the lap and any excess wiped off or it can be mixed with lapping oil to form a paste before it is applied. Vaseline and paraffin are other excellent lubricants that work very well on steel and iron. And machine oil is almost as good as any of these three.

Lapping paste (or valve-grinding paste), contains an abrasive powder that is suspended in thick oil or grease. It comes in five increasingly fine grades:

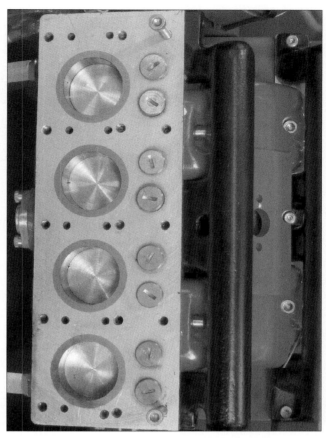

Photo 4.16: A four-cylinder internal-combustion engine needs its valves lapped into their seats and possibly its pistons lapped into their liners

1. Extra coarse – 30/60 grit.
2. Coarse – 80 grit.
3. Medium – 120 grit.
4. Standard – 180 grit.
5. Fine – 220 grit.

Choosing the right grade of grit depends whether the task involves rough, general or finish lapping. A fine abrasive will result in an excellent polished surface. For well-finished work pieces, only the two finest grades of lapping compound should be necessary.

Cleaning off the abrasive after lapping is essential. It is best done with paraffin using a brush, cotton wool or for any small holes, cotton-wool buds.

All abrasive materials will show signs of wear and their grains will break down into smaller pieces. At the start, the abrasive grains are brittle as well as hard and their sharp points tend to fracture under the stress of lapping. These fractures may leave sharp cutting edges or smoother edges, depending on the abrasive that has been chosen and the way it is being used. As all lapping

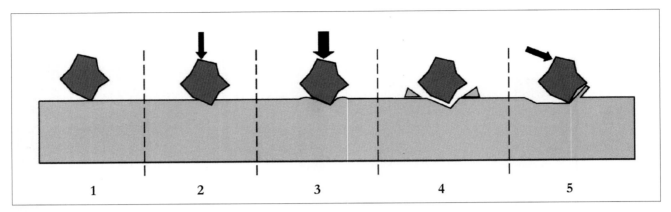

Drawing 4.17: The five stages in the lapping process showing how metal is removed

compounds are free to rotate, they will eventually end up round, having lost their ability to cut.

THE LAPPING PROCESS

To understand how lapping works, it is helpful to think about five separate stages of the process.

1. A grain of abrasive is sitting on the surface of a work piece.
2. If a downward force is applied, there will be stress at the contact point. As the abrasive is harder than the metal, the grain will break into its surface. With a minimal amount of pressure, any surface deformation will be temporary.
3. Increasing the pressure will force the grain further into the metal. If it is ductile, plastic flow will occur.
4. If it is brittle, one of more chips will form.
5. By moving the abrasive sideways, with sufficient pressure it will form a groove in the metal, displacing it until first one chip is formed, then another.

At the same time, heat will be generated both by friction and the deformation of the metal. These five stages are illustrated in Drawing 4.17.

In fact, the grains of abrasive material are not entirely free standing but are embedded in the lap, which must be a softer material than the metal being lapped. Otherwise, the abrasive will charge the work piece and cut the lap, instead of the other way round. As the pressure on the grains mounts, the depth they penetrate into the metal increases, assuming the grain size is large enough. Thus as the size of the chips rises the amount of metal they remove increases. However, the lapping process itself will cause the large grains to fracture, reducing them to smaller grains over a period of time.

As mentioned, lapping is divided into two categories. The first is equalising lapping that is used to create or improve the fit between pairs of a model's components such as pistons in cylinders and valves on their seats. Lapping the two parts will usually make the fit between them better provided they have first had their surface irregularities removed; usually by honing them before commencing lapping.

The other is called form lapping that aims to produce components of precise shape or size (flatness or roundness, length or diameter). For flat surfaces, this is done using an accurate flat surface and transferring it to the work. For cylindrical ones, it is done with a lap that has been carefully and correctly shaped and sized.

The best finishes will come from lapping both surfaces together using a series of abrasives that are progressively finer. Depending on the smoothness required, lapping should be done with several lapping compounds each with successively finer grits until the desired result is achieved. However, all the coarser lapping grit must be removed from both the lap and the work piece before a finer grade is employed.

Circular components

Internal and external cylindrical lapping may be carried out with a cylindrical lap in a lathe. Separately lapping both shafts and their plain bearings can produce very accurate sizes with a quality surface finish. But do not try and lap a shaft directly into its bearing. The wear rate of the hard shaft during lapping will be lower than that of the softer bearing. And should the tougher part be less well finished, the softer one will adopt the less precise shape. Additionally, there is a danger that grains of abrasive will become embedded in the soft bearing material.

Lapping pistons or piston rings into cylinders, as already stated, is not recommended until they have first been honed, since any irregularities in the surface of the harder material will simply be transferred onto the softer one.

However, once honed, lapping can be undertaken to provide as close to a perfect fit as is possible. The rate of rotation should aim to give a surface speed of approximately 1,500 surface metres per minute (5,000 ft/min), subject to practical experimentation with individual metals, i.e. speeds of 750rpm for a diameter of 6mm (0.25"), 300rpm for 18mm (0.75") diameter and 200 rpm for a diameter of 25mm (1"). A fine finish is particularly needed where O-rings rather than metal ones are used.

For lapping round holes, the internal lap needs to be a metal cylinder that perfectly matches the internal shape of the part and of marginally smaller diameter to make it possible for the lapping compound to fit in between the lap and the hollow part's wall. The lap itself must be softer than the metal of the hole's walls so that grains of the lapping compound bed in the lap, not the wall.

External laps are used to improve the outside diameter of cylindrical components and also to dress internal laps; but remember they must be of a softer metal than the internal lap being dressed.

Form lapping, properly done, does not significantly increase the diameter of a hole. Where the hole is a very slight interference fit (inserting the lap needs force), lapping may change that to a precision sliding fit. Bigger hole-size changes are best done first with a reamer or a careful boring operation. Laps that are made of a fine-grain iron, with its porous structure, can be purchased with straight or spiral expansion slots.

For internal lapping, mount the lap arbor in a chuck and then slide the lap onto the arbor, large internal diameter first so that the tapers mate. Then tap the lap until it is firmly on the arbor when its diameter will be as specified. Apply a thin layer of lapping compound and, to size the lap to the work piece, move it onto the lap and start to expand the lap until a light interference fit is achieved. With the chuck rotating, move the work piece back and forth over the lap, moving it so that the amount of lap visible at each end of the work piece is equivalent to about one-third of the work-piece length. Every ten seconds or so, remove the work piece from the lap, turn off the lathe, wipe any existing abrasive from the lap and replace it with fresh lapping compound. The correct amount of compound keeps the lap moist yet avoids compound collecting at the ends of the lap. Excessive application is likely to produce a bell-mouthed result. It may be necessary to expand the lap to compensate for any wear, or to increase marginally the size of the hole in the work piece.

Lapping the external diameter of a part follows the same process except that it is normal to fit the work piece in the chuck and the lap and its holder in the tail stock. The external lap is fitted in its holder and the screws should be adjusted to prevent rotation and to alter its diameter. The lapping compound is usually applied to the work piece, rather than the lap.

Flat pieces

Lapping can be used to produce components that are flat to fine tolerances and often with a polished surface finish. However, a flat cast-iron lapping plate will be needed, or a rotating one although the latter are seldom used by model engineers. A hand-lapping plate with slots or grooves in the surface will minimise the amount of work needed.

The part being lapped should be moved in a figure-of-eight pattern to avoid the work ending up with raised edges. One sign of the accuracy of lapping and the resulting surface finish is that quality flat-lapped surfaces tend to cling together when placed in contact and are hard to separate. However, this degree of accuracy is unlikely to be applicable to most model-engineering needs.

NOTES

5: REAMING

When the finish within a drilled hole is insufficiently good, a solution to improving it lies in reaming out the hole. Reamers are most likely to be employed in the home workshop to create circular plain bearings in which shafts or axles will turn. They may also be used where a press or interference fit is needed. All these tasks are feasible because a reamer is a precision cutter that completes a pre-drilled hole to an accurate size and circularity, together with a first-class surface finish.

However, a reamer will follow any misalignment exactly if the original hole was crooked. Thus reaming cannot be relied on to correct the location or the alignment of a hole. To give an example, the allowable tolerance on the diameter of a 12.5mm (0.5") reamer is only 0.075mm (0.0003"). And reaming is not a heavy cutting process. The walls of a reamed hole will certainly be smoother than those of a hole that has been drilled or bored. A reamer does this by removing a minimum amount of metal from the sides of the hole as it rotates in the same direction as a twist drill but at a slower speed; typically half the speed but with double the feed rate. A similar type of coolant to that needed for drilling, as well as an adequate supply, is essential to achieve a good surface finish. However, as technology has moved forward, a few carbide drills can provide a similar degree of precision to reamers.

Since reaming removes very little metal and is a delicate process, it may be carried out by hand or in a lathe. The latter method provides more axially accurate holes. Reamers should never be used to enlarge holes by a significant amount (more than around 5%). The best type of reamer for any particular task will depend on the material to be reamed, the size and depth of the hole as well as whether it is blind or not. The surface finish and the accuracy of the size needed are also important. These parameters will impact on the size of the reamer, its material and its type. A reamer is a precision tool. Successful use of any reamer depends as much on the way it is used and the machinery employed as it does on the reamer itself.

REAMER TYPES

Reamers come in a range of sizes from 1.5mm (0.06") upwards; unfortunately imperial sizes can be as much as 50% more expensive than metric equivalents. It is also worth remembering that brand-new reamers tend to cut fractionally over size. Reamers are multi-fluted with a number of cutting edges around a central shaft and some look similar to milling bits. This is illustrated in Photo 5.1. Reamers are made from several different materials of which both steel- and carbide-tipped ones are widespread and affordable. There are also many different types, each one designed with a particular function in mind. The only ones likely to be found in the average home workshop are one-piece hand and machine reamers, adjustable hand reamers and finally tapered reamers.

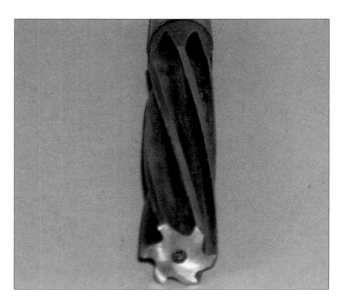

Photo 5.1: A machine reamer showing its six cutting edges

CHAPTER 5: REAMING

Photo 5.2: From left to right, two machine reamers one with curved flutes the other with straight ones, two hand reamers also one with curved and one with straight flutes, and a machine reamer with a morse taper and flat

Occasionally, it will be essential that the reamer is long enough to deal with a lengthy drilled hole, particularly if it has been created with an extra-long drill. As with drills, long-version reamers are also widely available but tend to be expensive.

A typical reamer comprises a set of parallel-helical or straight cutting edges that run along about half the length of its cylindrical body. Those with spiral flutes have their spiral in the reverse direction to those of a twist drill. This forces the swarf away from the cutting edge to avoid spoiling the hole's finish. A standard reamer thus has a left-hand spiral and is used for right-hand cutting.

The grooves in the body of a reamer ease chip clearance and enable cutting fluid to reach the cutting edges. The reamer is held and driven by its shank when reaming a hole whether by hand or by machine. The angular cutting portion at the other end is bevelled to help the reamer self-centre as it enters the hole. There are three different types of reamer, the first two for machine use, the third for hand reaming:

1. Cylindrically-ground parallel-shank without a square but sometimes with a flat for a retaining screw.

2. Standard taper shank for operating in a Morse-taper tailstock.
3. Parallel-shank with a square at the end for fitting in a wrench.

The cutting edges of reamers are ground at a slight angle and with a small degree of undercut below them. To avoid vibration and to prevent chatter marks, hand and machine reamers are manufactured with an even number of teeth and an uneven pitch (differential pitch). The difference in tooth pitch is 4° to −2° with a small number of teeth; 2° to −30° with a large number. The change in pitch largely avoids the above mentioned difficulties. The differential pitch is so designed that two teeth are always opposite one another.

The cutting-end chamfer of a reamer is usually around 45° on machine reamers; less on hand reamers. Along the flute length a cylindrical-guiding land is ground followed by a secondary-clearance angle. This land is crucial to guiding the reamer and sizing the hole it cuts. This is why clearance does not extend right to the edge; a cylindrical-ground land of a few hundredths of a millimetre (thousands of an inch) is left. The edge on the cutting chamfer is ground to a point.

METAL FINISHING TECHNIQUES ■ 57

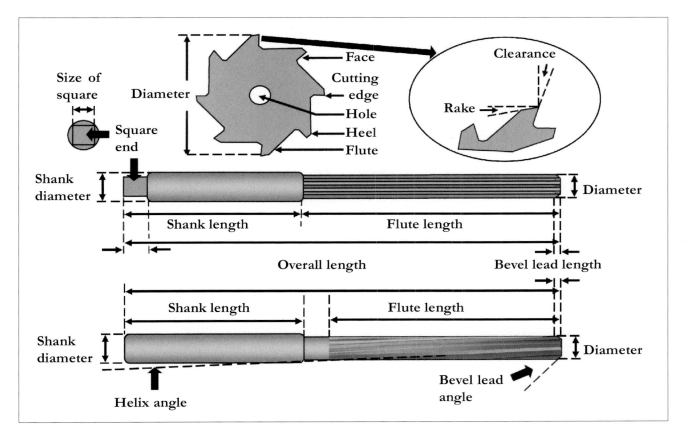

Drawing 5.3: The cross-section of the cutting faces of a reamer and the parts, top, of a hand reamer and, bottom, a machine reamer

HAND REAMERS

Hand reamers always have a square-ended shank for hand-driving using a tap wrench. They also have a slight further taper lead behind the 45° cutting bevel to assist the reamer in centralising itself in the hole should there be any misalignment. It also makes the start of the cutting action easier to achieve. Usually, hand reamers have straight flutes but some are manufactured with curved ones. The 45° bevel lead at the tip of a reamer is its cutting edge while the lands guide and centralise the reamer in the hole. Never try and use a hand reamer in a machine tool as power operation can damage the reamer and the work piece.

MACHINE REAMERS

Machine reamers have either a parallel or a Morse taper shank; the latter occasionally with a flat for a set screw in a tool holder. They never have a square end. The main types of machine reamers are solid reamers made from high-speed steel, carbide, or steel with brazed-carbide tips. Two other types of reamer are those made with replaceable carbide heads and blade reamers comprising detachable inserts and guide pads, which improve the surface condition by burnishing after the cutting part of the reamer has passed through the hole. As with many other tools, diamond reamers are widely available with specifications beyond the needs of model engineers, and at unaffordable prices.

Reamers for use in the home workshop are mostly manufactured from high-speed steel or alloy steel with tungsten-carbide cutting edges (coated carbide, cermet, diamond and CBN are used industrially to improve life, particularly for high-speed reaming). They have chamfered ends that cut the sides of the hole being reamed and the standard 45° lead-in angle is generally effective. Flutes may be straight, with a right-hand or a with a left-hand spiral. Straight flutes are for use on metals that do not form chips, such as cast iron, bronze and free-cutting brass. Do not use straight-flute reamers on holes with any intrusion, such as a cross-hole or keyway.

Left-hand spiral reamers (rotate clockwise when viewed from the cutting end) also work well on hard metals and will push any chips ahead of the reamer; thus requiring an open-ended hole. They must never be used on blind ones. A left-hand spiral provides a negative cutting action and attempts to push the tool out of the hole. This type of reamer cuts well to size, provides a good finish and is effective at bridging any interruptions in

Photo 5.4: A set of adjustable hand reamers that can between them cover a wide range of different hole sizes

the hole. The less common right-hand spiral reamers (rotate anti-clockwise when viewed from the cutting end) work well on hard materials. They pull any chips out of blind-holes and provide a positive cutting action, which draws the tool into the hole. However, a right-hand spiral may cut slightly oversize.

ADJUSTABLE HAND REAMERS

An adjustable hand reamer is capable of reaming a series of holes over a very limited range of sizes, with typically only a millimetre or two (forty to eighty thou) of adjustment in model-engineering sizes. A reamer's size may be measured across two opposite blades with a micrometer. Adjustment employs two screwed collars which lift or lower the replaceable cutting blades by movement along tapered seats. They can be locked in place at the chosen diameter. Do not over tighten the nuts as the slots and threads can easily be damaged. The fact that adjustable reamers do not have spiral flutes means that too heavy a cut is likely to cause them to chatter.

EXPANDABLE MACHINE REAMERS

Volume runs on abrasive materials require the use of expandable machine reamers that can be enlarged for regrinding. However, the adjustment is not designed to ream holes in a number of varying sizes. An expandable reamer may cut an incorrectly sized hole if the reamer is set to the wrong diameter, if it is worn or if material is

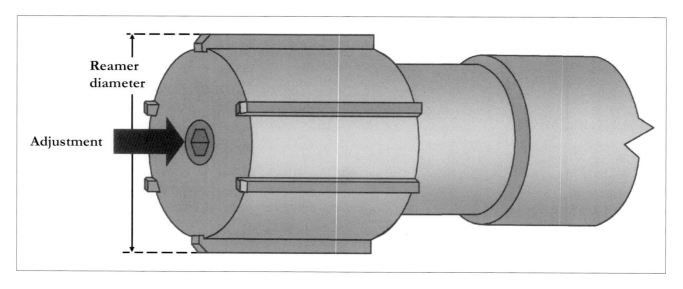

Drawing 5.5: An expandable machine reamer showing its Allen-key adjusting screw

METAL FINISHING TECHNIQUES

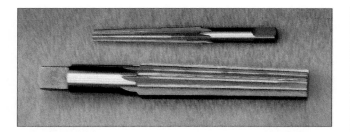

Figure 5.6: A pair of morse-taper hand reamers

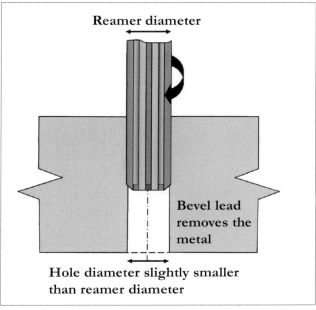

Drawing 5.8: The hole must be fractionally smaller than the reamer

allowed to build up on the cutting edges. A burr at the start of a hole or a tapered hole both result from poor alignment. A crooked hole can be the result of both concentricity and alignment errors or sometimes an inadequate lead into the hole. A poor finish is most likely to result from one or more chipped cutting edges, chips that fail to come out of the hole, incorrect grinding or excessive misalignment. Needing too much torque to turn the reamer may indicate a regrind is required, the reamer is set to remove too much metal or inadequate coolant.

TAPER REAMERS

Unsurprisingly, precision taper reamers are designed to finish tapered holes to size. Taper-pin reamers are ideal when tapered pins are to be used. In this case, it is recommended that the hole is drilled slightly smaller than the thin end of the taper pin. A taper reamer is defined by the diameter of its large end and the taper ratio, expressed as the diameter reduction per unit length or the included taper angle. They are made with a taper of 1:50 for metric pins or 1:48 for imperial-sized tapered pins. Hand-operated Morse-taper reamers have a taper of 5.21mm per 100mm (0.625" per foot) and are used to finish Morse-taper housings and sleeves.

Some taper reamers are hardly precision tools. They are specifically made just to de-burr drilled holes or even to enlarge them in thin sheet. One is illustrated in Photo 5.7 and is particularly suitable for working softer metals like aluminium, brass and bronze.

USING REAMERS

Reaming is a precision operation, whether carried out by hand or for preference in a lathe. Hand reamers may be used to finish holes manually but care is needed to ensure that the reamer is aligned with the hole at entry and is pressed in to start the cutting process. Begin turning and apply even pressure while turning. A cutting fluid will improve the surface finish of the hole and extend the reamer's life. Never turn a reamer backwards even when removing it from the hole.

Machine reamers will usually be fitted in a lathe chuck. Ensure that the work piece is held firmly and the right speed is used; a surface speed of around 7 – 15 metres (25 – 50 feet) per minute for mild steel, half that speed for high-strength alloy steels, 50% faster for brass, bronze and copper and twice as fast for softer metals like aluminium alloys and cast iron. The reamer should be fed into the hole at roughly twice the rate used for drilling the same metal. Too low a feed rate may reduce the finish and roundness of the hole and also increase reamer wear. Aim to pass the reamer just once through the hole, always turning it in the correct direction to avoid spoiling the finish of the hole and, at the same time, rapidly blunting the reamer.

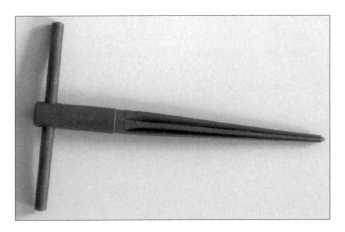

Figure 5.7: A tapered reamer that is useful for de-burring and enlarging holes in thin metal

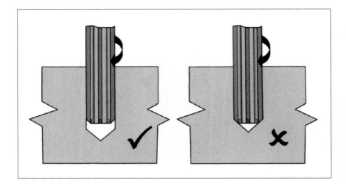

Drawing 5.9: As with drilling, clearance for chips is needed at the bottom of a reamed hole

The amount of metal to be removed by a reamer will vary depending on the type of reamer, the size of the hole in relation to the size of the reamer and the metal to be reamed; the softer the metal, the more a reamer can remove. The size of the drilled hole in relation to the size of the reamer is critical. If too little metal is to be removed, the reamer will rub rather than cut. However, no reamer can remove metal from a hole smaller than the inner diameter of the cutting edge. It is thus important to ensure that the drilled hole to be reamed has sufficient material to enable the reamer to cut it rather than burnish it. However, if the reamer has to remove too much metal it will soon wear and end up producing under-size holes; definitely not what is wanted! Holes of the following diameters should be undersize by the indicated amounts:

Up to 6mm (0.25") – 0.25mm (0.01").
6mm to 12mm (0.25" to 0.5") – 0.4mm (0.015").
12 to 25mm (0.5" to 1.0") – 0.5mm (0.02").

When using an HSS reamer in a machine tool, always apply plenty of cutting fluid and set the speed to about two-thirds of that for the drill that made the hole. And remember to remove the reamer from the hole before stopping the rotation.

To produce quality machine-reamed holes, ensure the work piece is securely mounted, the speed is correct, the reamer and the drilled hole are aligned and the latter is correctly sized in relation to the reamer's dimension.

Photo 5.10: An oscillating steam-engine cylinder in the 4-jaw chuck ready to be hand reamed by turning the lathe headstock with a handle

Keep reamers sharp by regrinding just the bevel and taper leads whenever necessary and ensure the flutes of a reamer do not become choked with swarf. This is particularly important in the case of any straight-flute hand or machine reamers.

A number of more common reaming problems that are likely to occur will spoil the finished hole:

1. A poor surface finish can result from too high a speed or feed rate, vibration or chatter, a lack of lubrication, a worn reamer, trying to remove too much metal from an undersize hole or a burred hole.
2. Bell-mouthed holes can occur if the feed rate is too low, there is vibration or chatter, shortage of lubricant or the reamer is worn or bent.
3. An oversize hole can be caused by too rapid a speed or feed rate, vibration or chatter, a lack of lubrication, a worn, bent or misaligned reamer, having to remove too much metal from an undersize hole or a hole that is burred.
4. An undersize hole is normally the result of a worn reamer but can also be caused by an inadequate amount of lubrication during reaming.

Thus it is essential to keep reamers sharp and avoid any damage to them. In addition, a correctly sized hole that is burr-free is a key requisite.

6: BROACHING

The technique of broaching may be used to create holes with shapes other than round – square, rectangular, D-shaped, triangular, hexagonal – as well as to cut keyways in metal. It is possible to broach any metal that can be turned or milled including castings, whether ferrous or non-ferrous. While similar work can frequently be produced with a file, the accuracy and quality of finish produced by broaching are both infinitely superior. The process can often be completed with a single pass of the broach and using such a tool may be the only way to produce internal straight-sided profiles.

Correctly used, a broach can achieve very tight tolerances and create holes with an excellent finish without much practice. However, having a precisely sized and shaped tool is the key to successful broaching. The tool can combine teeth from roughing through to finish cutting in a single tool regardless of its shape and its size.

Limitations of the method include the necessity for exactly the right shape and size of broach, the need to drill a pilot hole, the requirement for no obstructions in the tool's path and problems when broaching tapered holes.

Broaching involves removing material by either pushing or pulling the broach through or past a component. The teeth on the broach are equally spaced, with each tooth successively larger and closer to the final shape desired. Each tooth will only remove a few hundredths of a millimetre (thousandths of an inch) of metal, until the passage of the last tooth that gives the exact required shape and size. The action of broaching is similar to that employed by planers and shapers. The sharp cutting edge of the broach travels across the work piece and removes a specified amount of metal.

The process may be used on virtually all metals with the fastest material removal on softer ones like cast iron,

Photo 6.1: Cutting a key way in a flywheel is a classic application for broaching

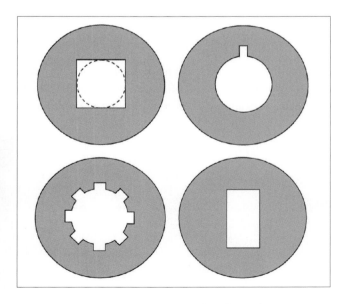

Drawing 6.2: Four useful applications for broaches; making square holes, cutting keyways, forming splines and creating rectangular holes

Photo 6.3: A selection of broaches for cutting keyways showing the variation in tooth length. Photo courtesy Home and Workshop Machinery

brass and bronze. Free-machining materials are easier to broach than hard ones. Steel is fine but the very hardest alloys will rapidly blunt the teeth of the broach. And the size of a broach can be as small as just under 1.5mm (0.05") to way beyond any size likely to be needed by a model engineer. The length of a broach is determined by the distance end to end of the metal to be removed and may be limited by the broach stiffness and the stroke of the machine holding it.

In the absence of a specialist broaching machine and because of access problems, broaches will normally be pushed into the metal rather than trying to pull them. And for strength reasons this does limit their minimum practical size.

The most common broaching operation in the home workshop is likely to be cutting an internal keyway in a gear or pulley, or forming the centre of a hand wheel to fit a square shaft. In addition, clock makers can use broaches to obtain square holes in the hands of clocks.

The main consideration with broaching is that each individual keyway, square- or other-shaped hole is likely to require a broach of different profile and dimensions.

A wide range of broaches with many alternative shapes and sizes are commercially available. However, they can readily be made from silver steel although a fair amount of work is involved in cutting the teeth and hardening them. Drawing 6.5 illustrates the various parts of a broaching tool.

TYPES OF BROACH

There are two different sorts of broach; internal and surface ones. Both can include roughing, semi-finishing and finishing all in one tool; typically a long, tapered, multi-toothed bar, rod or plate with teeth. When forced through a hole or moved across a surface, it cuts to the desired shape or size. Some broaches are even fitted with burnishing sections for further improvement of the finish (Chapter 7 gives more information about burnishing).

A broach consists of a number of progressively taller teeth cut on a single length of hardened steel, typically used to reshape a hole into a non-circular profile such as a rectangle or hexagon using a push or pull motion to remove metal. Model engineers may use a broach to cut square holes for components to fix onto square shafts, make keyways on shafts and the gears or pulleys fitted on them; less often to create inside splines in hubs. However, broaching can cut just about any profile, even involute gear teeth, only limited by the length and width of the shape to be cut.

Although broaches are commonly used to produce internal shapes and forms, external broaches are also employed to cut slots and other external shapes. It is possible to broach almost any shape of cross-section provided all its surfaces are parallel to the direction of travel of the broach. The amount of metal removed during the pass of a broach depends on the metal being cut, but is typically 0.05mm – 0.1mm (0.002" – 0.004").

Photo 6.4: A set of keyway broaches, slotted bushes and shims for varying the depth of cut

should be off and the chuck or faceplate locked in position to stop any rotation.

Similar techniques may be employed to cut square, rectangular, D-shaped, triangular or hexagonal holes with appropriately shaped broaches. In these cases, place the end of the broach in the component's pre-drilled hole, ensuring that the part is securely supported and then carefully push the broach through.

In industry, special broaching machines are widely used but their single-purpose function makes them a rarity in home workshops. As an example, rifling gun-barrels is an application that uses a very long broach with only a few teeth that can pass through the length of the barrel. The grooves it makes are only a few hundredths of a millimetre (thousandths of an inch) deep. A dedicated machine pulls the broach down the barrel at the same time rotating it to give the spiral pattern.

Another way of forming a different shape from a round hole does not involve any cutting process. A male form is forced through the hole in an annealed component; e.g. forcing a square shaft through a round hole to make corners that grip the square shank. This process only really works with a hard shaft and a relatively soft and

Photo 6.10: Forming a keyway in a lathe with a home-made broach mounted in the tool holder and a guide fitted in the flywheel's centre hole

well-annealed component. The method does not work with materials that tend to crack and it can distort the exterior of the component as metal is displaced rather than removed.

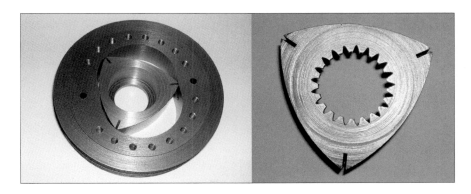

Photo 6.11: A Wankel rotor sitting in its housing, left, and a beautifully formed internal gear, right. Photos courtesy Warren Vickery

NOTES

7: BURNISHING & SCRAPING

BURNISHING

Burnishing is a cold-rolling process that uses a hard smooth tool, applied with plenty of pressure, to rub and stretch a metal surface without actually removing any metal. The pressure of the tool flattens the metal by compressing the minute high spots resulting in plastic flow of the metal.

Burnishing improves the surface of the metal by giving it a polished finish as well as enhancing its strength by work-hardening the metal. The result is a glossy surface, which is wear, fatigue and corrosion resistant. It is an entirely different process from polishing, which removes material in order to obtain the required appearance and produces a smooth finish, but not a hardened one. Polishing, as a first step, may be taken prior to burnishing to remove any significant surface imperfections. Shallow scratches can be repaired by using a burnisher to push metal back into the scratch itself, followed by sanding and then polishing to restore the surface.

Lubrication is crucial when burnishing; the oil should be easy to remove without leaving a residue, should hold tiny particles clear of the work piece and must minimise evaporation as a result of the heat generated by the burnishing process itself.

TYPES OF BURNISHER

The majority of mass-produced burnishers are made of quality hardened steel or carbide; occasionally sapphire or agate. An example of a steel burnisher is illustrated in Photo 7.2. Steel tools are readily resurfaced when they become worn; carbide ones are so hard they will rarely need resurfacing. To burnish clock pivots, both right- and left-hand burnishers are vital. The cross-section of each is a trapezium and when viewed from the end, right-hand ones tilt to the right and vice-versa for left-hand ones.

Burnishers can easily be made in the home workshop and there is no requirement for a cutting edge. Carbide ones must be finished with a diamond hone but hardened steel ones can be ground to shape and then polished. The alternative is to cut the burnisher to shape on a machine tool, then hardened and polished it.

Drawing 7.1: A magnified view showing that the burnishing process pushes metal from the peaks to fill the troughs

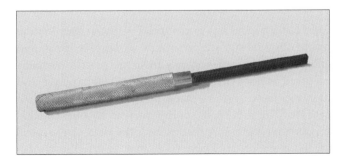

Photo 7.2: A classic commercial burnisher fitted with an easy-to-hold knurled handle

CHAPTER 7: BURNISHING & SCRAPING

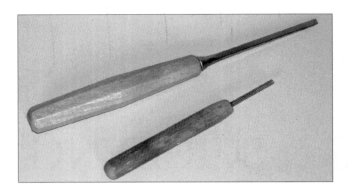

Photo 7.3: Two useful burnishers fitted with wooden handles

A pair of typical home-made burnishers that have been constructed from silver steel with wooden handles are shown in Photo 7.3. The metal should be ground to shape and smoothed with increasingly fine-grit emery until a polished finish is obtained. The tool should then be heated to cherry-red and quenched in oil to harden it. There is no need to temper the tip as a burnisher is not exposed to any shock loads and a high degree of hardness is the essential requisite of the tool. The final task is to polish the burnisher, preferably using a buffing wheel. Occasional re-polishing will be necessary to keep the burnisher in tip-top condition.

Burnishing can also be achieved using a wire brush, preferably one that is being rotated but even with a conventionally shaped brush worked by hand. Brushes are made with different degrees of firmness ranging from stiff to very soft. Stainless-steel brushes are corrosion-resistant and those employing fine-diameter wire are the best for burnishing without removing any of the base metal during the process.

Brass wire, softer than stainless steel, is still good for burnishing and also for polishing and de-burring. Phosphor bronze is stronger than brass and will last longer yet is able to undertake similar tasks. And fibre-glass burnishers can also often be used for hand-working metal.

Cold roller-burnishing works metal without either cutting or abrading its surface, improves the surface finish and enhances dimensional accuracy. Industrial practice normally involves roller-burnishing tools; several tapered and highly-polished rollers made from high-speed steel or carbide. The tool or the component to be burnished is rotated under pressure that exceeds the yield of the metal of the work-piece. These tools may be used in almost any rotating machine, but unfortunately, almost every application requires a different and expensive tool; not very practical in the home workshop. Furthermore, they

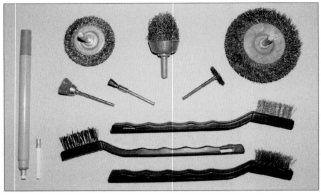

Photo 7.4: Three rotary brushes, three mini rotary brushes, three hand brushes and (left) a fibre-glass pen with spare refill can all be used for burnishing

can place excessive force on the bearings of the machine tool used, as well as on the cross-slide.

HAND BURNISHING

The most likely users of burnishing are those who work on clocks and watches, who will need to burnish the pivots in their mechanisms to minimise both friction and wear. The process improves the quality of the outer surface of a pivot by rubbing flat any tiny irregularities in the pivot's surface and work-hardening it. Grooves are quite commonplace in any worn pivots and they must first be flattened using a fine file or emery cloth. However, do not burnish plated pivots as the plating will flake and ruin the component.

For work on clock pivots, right- and left-hand burnishers have different profiles and methods of use. Right-hand tools have a relieved edge to enable them to provide a sharp transition linking the pivot and its shoulder. They are suitable for top use with pivots on their right sides and working their left ones from below. A left-hand burnisher is appropriate when working on the opposite edges. The choice of tool will depend on the handedness of the user as well as any preference for burnishing pivots from on top or underneath.

To burnish a pivot, the tool's surface must grip and stretch the pivot's metal. The burnisher should be flat and free of any surface defects. It should be prepared for work by moving the tool perpendicular to its length across fine crocus or emery cloth, placed on a really flat surface (metal or plate glass), or a flat emery stone. The aim is to give a series of shallow grooves across the burnisher's face. This task needs repeating at regular intervals. Any lines running along the length of the tool as a result of burnishing must be removed before continuing to avoid scoring the pivot's surface or damaging its finish.

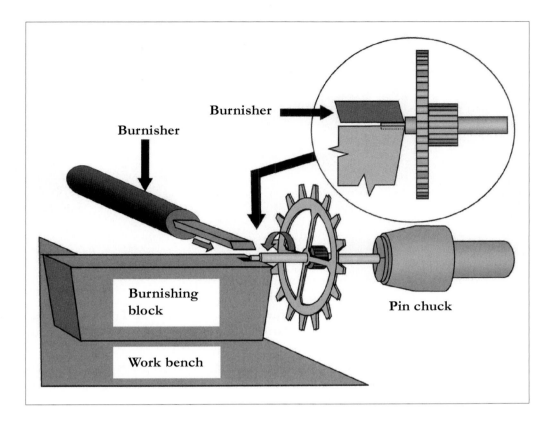

Drawing 7.5: A clock wheel and arbor with its pivot set up ready for burnishing

Using a pin vice to hold the arbor and wheel, support the pivot on a pre-prepared strip of hardwood that is around 20mm thick by 100mm x 45mm (a 4" length of prepared 2" x 1" wood) which has a series of grooves cut into it with a needle file to support various pivot sizes. The hardwood must be securely fixed to the workbench. Chamfer either or both the edges with the grooves in them to an angle of around ten degrees, as shown in Drawing 7.5, to clear the wheel. Before use, clean out the grooves with a thin strip of wood; a cocktail stick is ideal for this task.

Place the burnisher flat on the pivot that has been fitted in one of the grooves on the strip of wood. Holding the end of the pin vice with the thumb and index finger, rotate the arbor and wheel in a clockwise direction so that the top of the pivot moves in the opposite direction to the burnisher that is being pushed over it. It is essential to keep the pivot moving to avoid the burnisher producing flat spots. Repeat this process, regularly inspecting the pivot until satisfied with the resulting finish.

Hand burnishing is a still a relatively fast process despite the slow rotation speed of the pivot. Use some machine oil or paraffin as lubrication. To burnish a circular component successfully, such as a clock pivot, just the right combination of pressure and speed of pivot rotation are necessary, requiring both to be varied. Start with gentle pressure at quite a slow speed and increase both as burnishing proceeds. It is important to discover by trial and error the correct amount of pressure; too little will fail to burnish the pivot, too much may overheat or otherwise damage it. It is essential to keep any shoulder square to avoid it binding when the arbor and wheel are fitted to the clock.

When removing the burnisher from the pivot, move it away in a vertical direction to avoid damaging the pivot's surface. The pivot should be gently rubbed with small and subtle changes of pressure. The feel of the steel itself hardening and polishing becomes apparent with practice. Watching the places that lack any oil marks on the burnisher will show where the pressure is being applied. It is also possible to feel when things go wrong, like a fragment of metal trapped between the burnisher and the pivot that will scratch the finish. Always keep the burnisher clean and well lubricated to avoid small fragments of metal accumulating on it. The smooth, hard surface creates a long-lasting finish; desirable in clocks and watches since the key factor governing the life of a time-piece is the wear rate of its bearings. For any device required to run continuously, burnishing will significantly increase the time between overhauls.

BURNISHING IN A LATHE

The process for burnishing in the lathe is similar to doing the same task by hand; the lathe turning at low speed replacing the manual rotation of the pivot. The arbor

CHAPTER 7: BURNISHING & SCRAPING

Photo 7.6: A Jacot drum in a lathe tailstock with the arbor and wheel supported by a female centre with a drive plate and dog used to turn the arbor at slow speed. Photo courtesy Dushan Grujich

Photo 7.8: Tumble polishers to hold 680g (1.5lb) and 2kg (4.5lb). Photo courtesy W & W Co Inc

should be held in the lathe headstock but the free end must be supported either with a Jacot drum or with a custom-made brass support; a length of round stock with a hole drilled in its end. The top half of the resulting support must be removed so the pivot can be accessed.

To automate the process, roller burnishing can be used, where a precision-finished hard roller rotates against the work surface creating plastic deformation of the metal. A tool, such as the Morgan Pivot Polisher, can easily be fitted to almost any small lathe and employs a carbide roller that is independently rotated by a small electric motor in the opposite direction to the lathe chuck. It will burnish pivots rapidly and effectively, simplifying much of a demanding task that needs a lot of practice. As with all burnishing, adequate lubrication is essential.

Using the Morgan pivot polisher

Mount the gear which has a pivot to be burnished in the lathe headstock. Position the polishing wheel against the shoulder of the gear shaft to cover the pivot face. Adjust the pivot rest so the gear shaft runs true. Turn on the lathe and polisher and apply light to moderate pressure onto the top of the shield standoff. Light side pressure can also be applied for pivot shoulder polishing. Add a few drops of polishing oil to the pivot and oil wiper every few seconds. Carefully examine the pivot at frequent intervals with a high-magnification loup to check for a satisfactory finish which may only take a few seconds.

TUMBLE BURNISHING

A different method of burnishing can be applied to small components that can be rotated in a barrel with abrasive 'stones', all covered with burnishing fluid or water plus a few drops of domestic liquid soap. A range of geometric shapes and sizes of 'stones' are commonly made of ceramic, plastic or hard, polished steel. Typical turning speed is around twenty-five to fifty revolutions per minute, depending on the size of the barrel, for a few hours; the length of time depending on the type of metal being tumbled. The process also de-burrs metal. Small stone or pebble tumblers and polishers with a capacity of around a kilogram (two pounds) are ideal for tumble burnishing any small components. As an alternative, it is not difficult to build a motor-driven base to turn a commercial barrel.

Photo 7.7: The Morgan pivot-burnishing fixture. Photo courtesy The Morgan Clock Company

METAL FINISHING TECHNIQUES ■ 73

Photo 7.9: Myford lathe beds still have a quality hand-scraped surface

SCRAPING

Scraping is a method of producing a quality fit between flat or circular pairs of components. The technique of scraping was extensively employed in engineering until the mid-twentieth century. Early machine tools were largely made from cast iron. It was easy to form the complex shapes required, cut them with available tooling and hand fit by scraping. Two matching scraped cast-iron surfaces slide well when lubricated with a layer of oil. Hand-scraping became a well-established and skilled trade at a time when labour was relatively cheap. It was employed by machine-tool manufacturers for more than a century. It was also used when fitting white-metal bearings. However, once labour rates rose, attempts were made to find more economical solutions to the production of bearings and slide ways. Milling and grinding have both allowed manufacturers to produce excellent ways but fitting pairs of components together for a quality lathe or milling machine still requires hand scraping.

For skilled workers, the amount of metal removed by each pass of the scraper is only a few microns (less than a few hundred-thousandths of an inch). It is thus still possible to achieve a degree of flatness (or circularity) by hand that is unachievable by a machine.

TYPES OF SCRAPER

A scraper is a hardened hand tool that looks rather like a chisel. Photo 7.10 shows three examples of different types of scraper. The straight one with a marginally bowed end is for use on flat surfaces while the curved and triangular-shaped ones are for working on circular

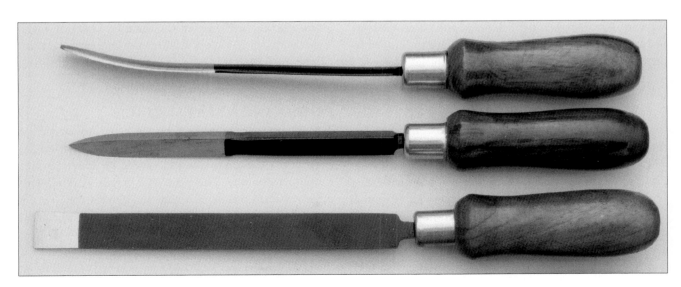

Photo 7.10: A curved scraper, a triangular one and a flat scraper with a slightly rounded cutting edge

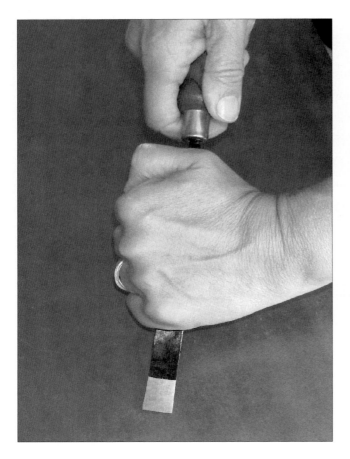

Photo 7.11: It is important that a hand scraper is held correctly to achieve a fine finish

bearings. Scrapers are used to transfer the flatness or circularity of a reference surface to a work piece. Normal use may involve creating a truly flat surface, fitting two flat surfaces together or matching a bearing to a shaft. This can all be achieved by marking the reference surface with some engineer's blue and then placing the work piece in contact so that the blue is transferred to the places where there is too much metal and some needs to be removed. In additional, hand scraping can be used to provide an attractive decorative finish on flat surfaces.

USING A SCRAPER
Before even thinking about using a scraper it is essential that its cutting edge is perfectly sharp. This may be carried out by honing the cutting edge to a fine polish. The right way to hold a scraper is demonstrated in Photo 7.11, making adjustments to the angle until the scraper is cutting efficiently. The process of scraping involves careful removal of minute amounts of metal where there are any blue marks. Scraping is quite easily carried out, even by inexperienced amateurs, to remove less than 0.0025mm (0.0001") and the process should then be repeated after re-bluing, working at right angles to the last direction of scraping until no blue is left.

The work piece will then match the reference piece. The quality of the resulting work is dependent on the skill and craftsmanship of the person who is doing the scraping; experienced workers will be able to produce an exceptional finish. The unskilled may, at the worst, produce an inferior surface to the one that existed before the scraping process started. Should a quality surface plate not be available, a sheet of thick plate glass or even a mirror can be used as a substitute flat reference surface.

Perhaps the most common use of scraping today is on the beds of machine tools. It is the sole way of achieving flat straight cast-iron surfaces that will consistently make contact across the whole of their mating surfaces. This ensures that slides do not strike the lathe bed causing the load to be concentrated at that point and the oil layer to collapse producing scoring, uneven wear or, in the worst case, particles of metal to break off.

The role of scraping is to try and spread the bearing contact evenly across the whole surface, while providing the right geometric alignment. A common standard used is 2 points per square centimetre (15 per square inch) of bearing area. Two or three times that number of points can be achieved with repeated fine scraping. The aim should be for a contact area that is over 50% of the total surface area. Spaces between high and low points act as oil reservoirs to minimise wear. Thus the life of a re-scraped machine-tool bed relies on the skill of the person doing the work. When the contact area exceeds 80%, the surfaces will try to stick together; not ideal for moving slides. And a 'perfectly' flat surface would rise slightly as it started to move over a thin film of oil.

Scraping leaves a distinct pattern on the surface finish, indicating precision ways. Their absence may signify an inferior product. However, the manufacturers of some low-cost machine-tools will add frosting to machine ways to give the impression of a well-scraped finish. Frosting is described in the next chapter.

PRODUCING A FLAT SURFACE
The ability to be able to produce a flat surface from scratch is a fundamental skill of scraping. Historically, apprentices learned to do this by using the 'three-surface rubbing' technique. This process involves working with three separate pieces of metal since, with only two, a pair that are not flat may fit together. Introducing a third avoids this. All three combinations of fitting the three surfaces together must be completed without any of them showing a lack of touching to prove that all three of the surfaces are truly flat. This is demonstrated in Drawing 7.12.

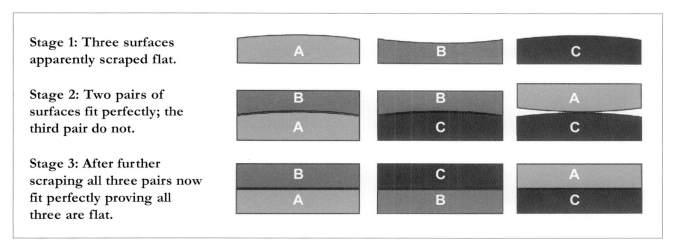

Drawing 7.12: Scraping a flat surface from scratch requires work on three separate work pieces

NOTES

8: BARE-METAL FINISHES

PLANISHING AND PEENING

Planishing aims to produce the final smooth surface finish to an already curved metal component by using a succession of rapid but light hammer blows. It is a process that removes any imperfections by hammering the metal between a well-polished hammer that is flat or slightly curved and a planishing stake. By applying repeated relatively soft blows the metal can be smoothed. The technique is known to have been used to make armour as early as the Bronze Age. The smoother the planished surface the less sanding and polishing will be needed and the lower the risk of thinning areas of the finished item. Running fingers over a planished surface is often a better way of finding areas that need more work than by just looking at them. In industry, planishing can also be achieved by blows from well-polished dies or by rolling in a planishing mill.

Peening, on the other hand, does not provide the smooth surface of planishing. Nor does it result in a fine gloss. However, peening also entails mechanical cold-working of metal, stretching it and coaxing it into shape by moderate repetitive hammer blows. Peening deforms the metal and tends to stretch its surface. It results in work hardening of the metal that makes surface cracking less likely to occur under fatigue conditions.

Model engineers are likely to have come across the process either when flanging the edges of copper boiler plates or when peening the ends of rivets to fix them in place and form the appropriate head or to fill the countersunk area. Shaping boiler plates involves first making a former from hardwood, medium-density fibreboard (MDF) or steel. The pre-annealed plate is then clamped in place and the edges hammered to shape; re-annealing at regular intervals to remove work-hardening. Peening can also be employed to remove a crease from folded or rolled metal sheet or to increase one dimension of a component that is marginally undersize.

Photo 8.1: The left-hand end of the shiny cover fitted over the dynamo's armature provides an excellent example of planishing

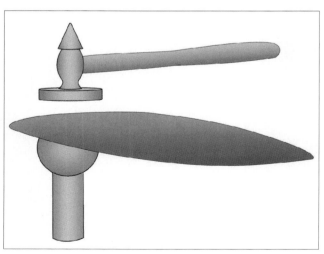

Drawing 8.2: Using a planishing hammer and post to curve a sheet of annealed metal

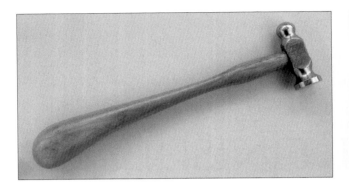

Photo 8.3: A small ball-pein hammer suitable for relatively delicate work

Always use a ball-pein hammer for peening as it has a harder head than a claw hammer. The process will also work harden the metal while expanding its surface and relieving tensile stresses, though it may induce compressive stresses.

Shot, flapper, glass-bead and laser peening are amongst techniques used in industry but are not really practical for model engineers.

CHASING AND REPOUSSÉ

Chasing and repoussé are very similar decorative metal-working techniques and are only likely to be carried out by model engineers who are building full-size or scale models where the original includes such ornamentation.

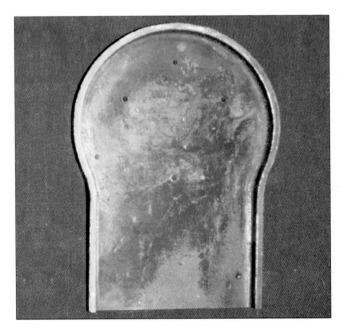

Photo 8.4: A peened copper end-plate for a boiler before drilling the required holes

Photo 8.5: A photo of a full-size traction engine in a classic repousséd metal frame

Probably watch makers and repairers are most liable to need to undertake this type of work.

Chasing is a metal-decorating method that uses a hammer and punch to produce a depressed design on the front of a piece of metal. Repoussé, on the other hand, is done from the back of the metal to provide a raised design. In both cases the metal is cold worked and is carried out with a range of suitable punches, a chasing hammer and a pitch pot.

The metal must be annealed to a very soft state. Both processes stretch the metal where it has been punched. The work is painstaking but can result in an eye-catching raised or depressed finish although the marks of the punches that have been used are generally visible.

Punches are normally custom-made for the task in hand from tool steel that is annealed, then cut, sawn and filed to give the tip a shape to create the appropriate pattern. The end edges of the tool should be bevelled before it is then tempered. Typical tools are liners that have a slightly rounded end, planishers with smooth, flat tips for pushing out sizeable areas of flat metal, tools with pattern ends to produce detailed areas and round or oval doming tools that create rounded areas.

CARRYING OUT THE WORK

Whether undertaking chasing or repoussé, it is first necessary to draw up the design pattern. Then transfer it

METAL FINISHING TECHNIQUES ■ 79

Photo 8.6: The signature of a famous British clock designer and maker engraved on a clock face

to a suitably sized piece of the chosen sheet metal that has been annealed and is flat and smooth so that the design is clearly visible.

The normal method of chasing or repoussé is to lay the thin sheet of metal to be worked on the surface of an iron bowl of warmed pitch which hardens when cooled to form a firm base on which to punch the desired pattern into the metal. The pitch can be heated in an oven or with hot air from an electric paint stripper. The pitch needs to be of the right consistency; not too hard or the metal will be thinned; not too soft or the metal will over-react when punched. The pitch should be soft enough to yield but sufficiently hard to maintain its shape. It can be made from a quantity of molten pitch (at around 175°C) into which an equal amount, or up to 25% more plaster of Paris is gradually mixed, followed by 7.5% linseed oil or tallow. A few companies will supply ready-made chaser's pitch. Take care to avoid being burned when working with hot pitch.

The annealed and marked-up sheet of metal is floated onto the surface of the molten pitch and pushed down until the pitch just reaches the upper surface of the metal but does not flow over it. When the pitch has set, work can commence punching out the pattern. Heat up the sheet of metal at regular intervals to enable it to be removed from the pitch, cleaned with turpentine and re-annealed as necessary before again floating it on the pitch and carrying out further work. The resulting patterns in the metal, whether raised or depressed, will then have to be thoroughly cleaned to complete the work.

ENGRAVING

An engraved surface will occasionally be needed by model engineers, particularly clock and watch makers, who wish to decorate their work. It may also be found on some musical instruments and almost every machine tool. Engraving may not be a finish in the conventional sense but is does provide a decorative appearance.

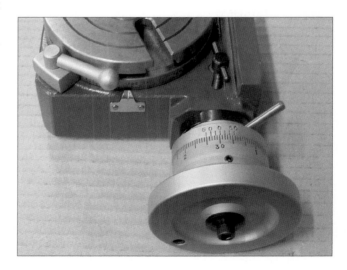

Photo 8.7: Not all engraving is decorative; here the rotary-table scale has engraved scales

CHAPTER 8: BARE-METAL FINISHES

Photo 8.8: A fine trio of gravers for cutting different patterns in metal

Hand engraving uses a sharp, hardened and suitably shaped steel tool or graver fitted with a small wooden handle that conveniently fits in the palm of the hand. This will cut into the surface of a sheet of metal using hard hand pressure or light blows from a small hammer. Some gravers are shown in Photo 8.7. The pattern to be engraved should be carefully marked out on the metal before work commences.

The graver's shape and its angle relative to the work surface affect the shape of the groove that it cuts. A square-profiled graver will cut a V-profile groove and is the most popular shape though there are lots of different alternatives. By holding the tool at the appropriate angles the thickness and depth of the cut furrow can be varied. When cutting a straight line, push the graver forward using hand pressure.

It is preferable when cutting a curved line to hold the graver still and feed the metal into the graver's tip. This requires a rotating vice or holding device to grip the piece of metal being engraved.

Engraving is a difficult technique to master without plenty of practice. The sharpening of the tip of a graver should present few problems to model engineers (See Photo 4.3 on page 45). What is required is the ability to grind and then hone the graver to the desired shape that gives an accurate, burr-free cutting edge. Burrs will be raised in the work piece especially when cutting curved lines if too much down force is applied or if the heel of the graver is too long or short. A long heel will generate additional drag but too short a heel will dig too deep. Both result in burrs along the edges of the cut.

A much easier but less traditional method is to use a powered engraver that may be an electric (hand-held or flexible-shaft driven) or a pneumatic engraving tool. These devices are increasingly popular. Both types are

Photo 8.9: A classic ornamental-turning lathe being demonstrated at an exhibition

Photo 8.10: A World War I aero engine surrounded by an engine-turned cowl

Photo 8.11: A fine example of engine turning ready to make a cowl for a small diesel engine

normally fitted with a carbide tip. Some use a rapidly turning electric motor. Others 'write like a pencil' with an oscillating carbide tip, even on metals as hard as steel. There is usually a calibrated stroke adjustment for different metals that alters the depth of engraving. This allows both fine lines and deep marks to be made. By carrying out the work in a lathe, it is then referred to as ornamental turning.

At the industrial level computer-controlled engraving enables the production of consistent and economically priced work. Similar results can be achieved in the home workshop with a computer-numerically-controlled (CNC) machine tool.

ORNAMENTAL TURNING

Ornamental-turning lathes were first produced in the fifteenth century but at first only for working wood and ivory. These machines consist of a relatively small lathe with an attached engraving or milling facility. They have the capability to position both the work piece and the cutting tool to follow paths that are not circular. There are many specialised chucks, cutters and other fixtures and fittings, often driven by an overhead shaft, that enable the most complex shapes to be produced. The most famous producer of these lathes was Holtzapffel during the nineteenth century.

A particular type of ornamental lathe is the rose engine that also dates back some five hundred years. These machines are quite rare today and antique ones are expensive to buy. They are mostly used by those who make jewellery.

The rose engine is also a lathe with a rocking and traversing headstock used for engraving wavy circular and elliptical lines. It utilises patterned discs or cams called rosettes. As the work piece is rotated, the mandrel rocks back and forth on the rosette while a tool cuts into its surface. Ornamental turning is a very specialised field that is a hobby in its own right.

ENGINE TURNING

Engine turning may be employed as a metal finish for one of two reasons. The first is to simulate a similar finish on the prototype being modelled. The second is for ornamental purposes on a custom-made model or full-size item such as a watch or clock.

Engine turning involves moving a metal work piece against a rigid tool in order to cut a series of regular patterns that result in a surface that produces interesting light reflections. For most applications the work involves using a specialist ornamental-turning tool; a straight-line engine or a rose engine for curves.

The lines can be parallel, elliptical or circular and may radiate from a point on or offset from the work piece. The surface of the metal to be worked should be as smooth, flat and clean as possible. If the engine-turned piece needs to be curved, it should be engine turned

Photo 8.12: A gas turbine fitted with a brushed-aluminium compressor cover

Photo 8.13: A brushed finish is a popular basis for some clock dials

first while the work piece is flat and then carefully bent to shape.

One easy way to produce an engine-turning effect that was popular on the cowls of World War 1 aero-engines is in a milling machine. Use a solid cylindrical piece of abrasive-impregnated, rubber-bonded material as the tool, rotating at around 1,000rpm; faster if necessary to get a better result. Decide on a spin pattern with overlapping swirls. Move the milling table a precise amount in the directions needed before each application of the rotating cylinder. To avoid over-heating the metal, use a lubricant such as WD40 or light machine oil and re-lubricate before each circle is made. Practice makes perfect so work first on a test piece of metal, noting the length of time the abrasive was applied to the metal and the amount of pressure used to accomplish the required appearance. The application of exactly these two parameter for each individual circular pattern is essential in order to obtain consistent results. Do not apply too much pressure as this will cause the end of the abrasive cylinder to spread resulting in ever-increasing circle diameters. As an alternative, a piece of dowel or even cork of a suitable diameter, mounted in the chuck with its end charged with carborundum paste can do the job.

BRUSHED FINISH

Although wire brushes are often associated with cleaning metal, a brushed metal finish is exactly that; a lightly-textured satin finish, usually obtained by applying a rotating metal brush to the component being worked.

As a process, wire brushing does not remove metal so it avoids clogging and will not change the dimensions of the work piece. It works better with softer metals than hard ones. Plain steel, stainless steel, brass and phosphor-bronze brushes can all be utilised. They are available in a range of hardness from very soft to stiff and in the form of wheels, cups, stem-mounted and tube brushes. Photo 7.4 on Page 70 shows some rotary brushes. Note that plain steel brushes can leave a residue that may rust.

Brushing pressure should be light and the speed fast; 650 – 800 surface metres per minute (6,500 – 8,000 ft/min). Remember not to exceed the maximum free speed of any particular brush. Soft brushes harden as speed increases so that, all other things being equal, a softer brush at a higher speed is preferable to a harder one at a lower speed. Excess pressure over-bends the wires of the brush, damaging them, generating unwanted heat and dulling the surface finish. Brushes may be fitted to a lathe or mill, or for very small components, to a miniature hand-held electric drill.

Alternatively, a belt sander will produce a similar finish as will a buffing machine with a suitably coarse abrasive. An alternative is the hand-use of wire wool which comes in a range of grades from 0000 (very fine) to 5 (coarse). And rubbing phosphoric acid into a steel surface with fine wire wool will produce a rust-resistant, silver-grey finish. Phosphoric acid is a greenish liquid or gel (often sold as a rust remover) that can be used on rusted iron or steel to convert the rust into a black compound.

Photo 8.14: An example of knurling a length of aluminium rod on a small lathe using a clamp knurler

KNURLING

Many knobs and tools possess a knurled handle to provide a firm yet comfortable grip for the user. There are two different ways of producing knurls; they may be cut or they may be rolled to deform the metal.

Cut knurls can be produced in any screw-cutting lathe with the same automatic-feed procedure that is employed for cutting screw threads; diamond knurling is, in effect, a series of left-hand and right-hand threads cut at a very coarse pitch.

Producing rolled knurls is carried out in the lathe with knurling wheels and requires the outside diameter of the work piece to be chosen so that the knurl wheels roll a whole number of patterns around the work piece. However, when working with soft metal, knurls have a high chance of success even with the wrong diameter. It is a good idea to do a test run on a spare piece of the same metal, turned to the same diameter, to ensure that a perfect knurl pattern can be produced on that size of work. If an overlapping pattern results, a small reduction or increase in diameter should resolve the problem and result in a perfect knurl pattern.

When cutting knurls, on the other hand, the spacing of the cuts is not fixed and can be altered to enable a whole number of patterns to be cut around the work regardless of its diameter.

Knurl rolling deforms the metal surface by pushing the cutting edges of the knurl into the work piece, displacing the metal until the complete pattern has been formed. Depending on the knurling wheels that are selected, the finish may be fine, medium or coarse and give a diamond, diagonal or straight pattern (see Drawing 8.15).

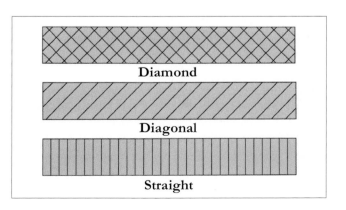

Drawing 8.15: Three popular knurling patterns; diamond, angled and straight

Photo 8.16: Above, a clamp-type knurling tool and below, a pair of knurls that will produce a diamond pattern

The process is normally carried out in a lathe using a knurling tool to roll a pattern in the work piece by impressing a series of ridges and grooves in the metal.

Roll knurling is best suited to the softer metals such as aluminium or brass; any attempt to knurl harder metals will almost certainly ruin the wheels. For occasional work, high-speed steel (HSS) knurling wheels are fine.

Two types of device for knurling are common. Both mount on the cross-slide of a lathe. The first employs pressure fed from the cross-slide while the second clamps the work piece from above and below. Both require plenty of cutting oil when forming the knurls and the regular clearance of the swarf.

The pressure-feed knurling tool, incorporating one or more knurls, is clamped to the cross slide and pushed against the work piece mounted in the chuck rotating at a very low speed. The pressure is gradually increased until the tool produces a pattern on the metal. Automatic traverse of the saddle is then engaged and as it moves along the lathe bed the knurled pattern is impressed into the metal along its length. Stop the lathe and check the knurl is deep, well cut and neat. Unfortunately, this type of knurling is not good for lathe headstock bearings.

Photo 8.17: Slight pitting caused by rust is already apparent on the barrel of this detailed model of an artillery piece

A clamp or straddle knurling tool is a much better device. The knurling tool is mounted on the cross-slide but the force is applied equally to both sides of the piece of metal being knurled, thus removing significant loads from the lathe itself. This type of tool is illustrated in Photo 8.16. The work piece is mounted in the chuck, the clamping nut tightened and the lathe set to rotate at a very low speed with auto-traverse or even turned manually using a handle engaged in the headstock. As knurling progresses, continue to increase the clamping pressure until the pattern is sharp, full depth and clean.

The final diameter of the work piece will be slightly larger than before the knurling commenced, due to metal displacement. This particular change in size can be used to enlarge a shaft diameter for a press fit by knurling in a straight pattern.

OIL OR GREASE COATING

Sometimes ferrous metal has to be left in a bare state. This may be because of the need for easy movement of closely fitting parts as in the case of a lathe bed. The other is to match the finish on a prototype or for an aesthetic purpose where a natural steel or cast-iron finish is needed. And unfortunately, steam, water and products of combustion may end up on the bare surface. A thin layer of oil or grease will stop the metal rusting.

In the case of new machine tools, particularly those shipped from the Far East, there will normally be a coat of protective grease to avoid corrosion during the voyage to the UK. This coating must be removed and replaced with a thin film of oil before the machine tool is used. This may be done by the supplier or the purchaser.

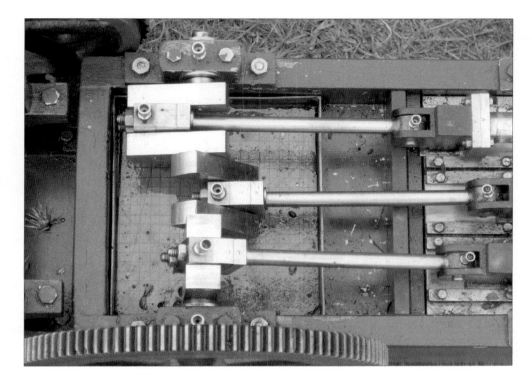

Photo 8.18: A fair amount of dirt will accumulate on any steam-powered model

FROSTING

Any oil film that is applied to a precision-ground surface may not be able to adhere, particularly where a pair of components that fit together have a quality finish. The film of oil will be removed as soon as any movement starts, exposing the bare metal and increasing the danger of damage to the surfaces and eventual seizure. As described in the previous chapter, scraping a surface will provide lots of tiny hollows that act as oil reservoirs. This final part of the scraping process, rather than full scraping to provide an accurate surface, is called frosting, flaking or spotting. The claimed benefit of frosting is that it allows smooth movement of the two surfaces, minimising friction by retaining oil between the surfaces, albeit with the penalty of a reduced bearing area. It is worth mentioning that some manufacturers of low-cost machine-tools add frosting to the ways of their machines to give the impression that they have a quality scraped finish even though this is not necessarily so.

OTHER RUST BARRIERS

Bare ferrous metals may be protected by applying a very thin water-repellent and corrosion-preventive film. The products for doing this either employ wax or sometimes grease dissolved in a fast-evaporating solvent. Solutions designed to give short-term protection between machining a part and finishing it generally use an oil-based product. Proprietary aerosols are widely available from specialist car repair and protection companies but care should be taken to ensure that the chosen product is completely transparent as some leave a coloured protective coat, usually brown. Most products are highly inflammable during application and can be removed with white spirit.

Of course, for storing any ferrous tools and parts in a home workshop, a piece of paper impregnated with a

Photo 8.19: Frosting shows up well on the left-hand side of the vertical slide of this low-cost milling machine

Photo 8.20: An attractive etched maker's plate on the front of a scale traction engine

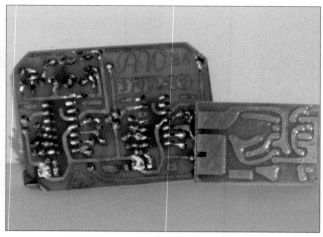

Photo 8.21: Home-etched printed-circuit boards, one with the components soldered in place

vapour-phase corrosion inhibitor (rust-inhibiting paper) may be used to wrap the item. This will avoid any need to apply a protective-surface coating that will later have to be cleaned off.

ETCHING

Model engineers may use etching in a number of different ways. The first is to etch letters into metal, an example being the etching of a model's name plate. A second is to etch entire small components that may be folded or soldered together. And it is the normal process for home-produced electronic printed-circuit boards.

The component to be etched can first to be covered with an etch-resistant material such as wax and the coating removed from the areas to be etched. A rather simpler alternative is to use press-down lines, pads and lettering, combined with an etch-resistant felt-tip pen. These can build up to the required layout. Next the item must be immersed in a suitable etchant in a plastic tank; ferric chloride for copper, dilute nitric acid for brass and copper, dilute hydrochloric acid for aluminium and dilute sulphuric acid for steel.

When a sufficient depth of metal has been dissolved from the component being etched, it should be washed clean under running water which, together with the etchant, must be disposed of with care; not down the drain. Press-down lettering and felt-tip pen ink can be removed with a suitable solvent like acetone. If the etch-resistant material is a wax it can be melted off the component by the gentle application of heat, remembering that hot wax is inflammable and can cause nasty burns. Finally, the etched work should be given a good clean and, if not a printed-circuit board, a polish as well.

9: METAL COLOURING

There are several methods of colouring a metal surface without painting it, such as anodising, plating or patinating. Some of these colouring techniques simply involve heat treatment, others use either chemical or electro-chemical processes. None described use paint, varnish or lacquer which are covered in the next chapter.

Before starting any colouring process, the metal must be absolutely dry and clean; oil, grease and polish free. Some techniques are easy to use in the home workshop; others need unpleasant chemicals, often heated, making them even more objectionable processes. Only potential users can decide whether to use some of the more hazardous processes described below. If they do, then care must be taken not to endanger themselves, their families or the environment.

BLUING OR BLACKING

The terms bluing and blacking refer to the same process, named after the blue-black appearance of the resulting protective finish. It was probably significantly employed first on guns to provide a protecting layer. The process is popular with clock and scientific-instrument makers, and restorers, for bluing small bolts as well as clock, watch and instrument hands. Steel and iron can be blued in three different ways but some alloys do not blue as well as others, so it is always worth testing a sample before making the component.

1. Use heat to temper the metal until it turns blue (it is the colour of the metal rather than its temper that is of interest) when a surface coating of black oxide (magnetite or Fe_3O_4) is formed. This method will provide a significant degree of rust and corrosion protection but only if the blued component is given a thin coat of oil or wax that also emphasises the colour. The process does not add any appreciable thickness to the component and thus does not affect the fit of blued components. The surface oxide is so thin that the component size will not change significantly; no more than 0.0025mm (0.0001").
2. Immerse the part in a heated chemical solution to provide a similar oxide coating.
3. Employ a cold process that uses one of a number of proprietary blue products (most are based on selenium compounds).

HOT BLUING

Hot bluing by heating in air is well suited to small components as it is essential that they are evenly heated throughout to around 300°C; the particular temperature depending on the exact shade of blue that is required. In all cases, before the bluing process can start, it is vital that the components are clean, polished and then have any oily or greasy deposits, including any finger marks, removed from all their surfaces. This can be done with alcohol, methylated spirits or any other volatile solvent.

A practical way to blue small items is to heat them (but not over-heat them) on a bed of fine sand (bird sand

Photo 9.1: Elegantly shaped blued hands as well as blued bolts can be seen on this unusual clock

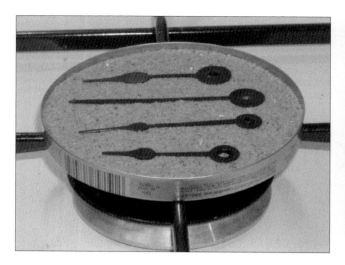

Photo 9.2: Blackening clock hands in a tin lid of bird sand on a gas hob

sold in pet shops is ideal) around 3mm (0.125") deep until the required colour is reached. The heat must be evenly applied to the bottom of the container to give consistent bluing of the components on the sand bed. Let them cool in clean oil or water (which will then need to be removed).

An alternative is to use an electric hot-air paint stripper with the component placed on a heat-resistant board. Altering the distance of the gun from the work piece is the best way to control the temperature.

Other methods of hot bluing involve heating unpleasant, corrosive chemicals well above the boiling point of water. Adding water to the hot mixture to replace the rapid evaporation losses can result in the concoction turning explosively to steam and spraying the very hot and corrosive liquid over anyone and anything nearby.

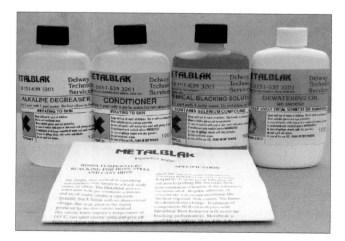

Photo 9.3: The four chemicals and instructions supplied by Metalblak

Photo 9.4: A bracket attaching a digital readout to a lathe saddle, darkened with Metalblak

Tectrate steel-blackening salts provide a repeatable solid black finish on most steel and iron. Pre-polished surfaces take on a deep blue/black colour. The finish provides corrosion protection and abrasion resistance. However, the chemicals are corrosive, irritant, flammable and the work has to be carried out at 141°C in a steel container. This method of hot bluing is unsuited to home use and the hazardous waste will need to be safely disposed.

A safer but messy and smoky alternative is to heat steel parts in an oil bath. Unfortunately, this process also involves a significant fire risk.

COLD BLUING

Cold bluing is not particularly resistant to wear but is a good way of preventing rusting. It involves treating the surface of the metal with a bluing fluid of which many proprietary brands are available such as Carr's Metal Black, Metalblak and Koolblak.

Polish the components to remove all tool marks and ensure that any polish, oil and grease are removed. Then immerse them in the bluing solution and they will quickly blacken. Remove and rinse in hot water, allow them to dry and carefully polish.

Metalblak (and Koolblak) steel-blackening salts come in a kit that includes all the chemicals for degreasing, conditioning and blackening, a protective sealing oil and full instructions for use. After degreasing, rinsing, conditioning and rinsing again, immerse the item to be coloured in the blackening solution diluted with three

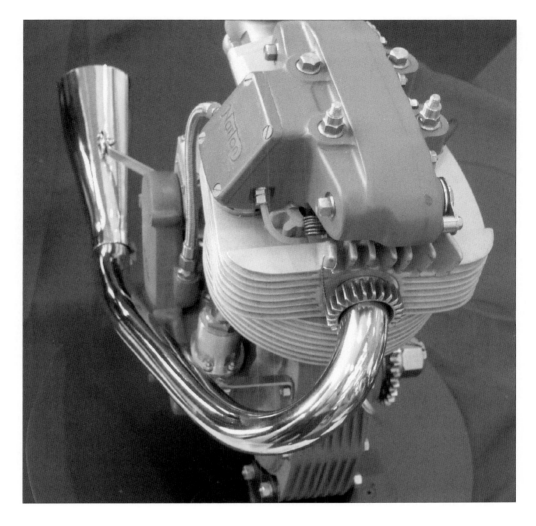

Photo 9.5: A beautifully formed, chrome-plated exhaust pipe; essential for a realistic scale appearance

times the volume of tap water at room temperature for one to two minutes. This will give a non-electrolytic, permanent blackening to steel and iron without affecting the component strength or changing its dimensions. After rinsing again, immerse in the sealing oil for ten minutes and allow to dry. The chemicals may be held in a plastic container and sealed for re-use.

Liberon Haematite will blue bright iron and steel once the surfaces have been thoroughly cleaned to remove any residual grease or oil. Ensure the metal is dry before applying Haematite either with a brush or a cloth until the desired colour is achieved and then wipe dry. Alternatively, the items may be immersed for a maximum of two minutes in nine parts of water to one of Haematite to ensure a uniform colour change. When the desired colour is achieved remove from the liquid and wipe dry. With both processes, after drying immediately apply a water-displacing oil like Liberon Jade Oil to fix the colour and to prevent any further oxidation taking place.

Solutions are also available for blackening most other metals. Carr's produce a range of their Metal Blacks that are suitable for use on aluminium, brass, nickel silver, solder, steel and white metal.

Some bluing chemicals contain selenium. They are safe to use cold but never heat them. Also do not touch, inhale or swallow liquids containing selenium.

ELECTROPLATING

Electroplating is a particularly demanding process to undertake in the average home workshop. However, to protect any metal which may corrode, plating with a corrosion-free metal can also provide a decorative finish that will be a match to the prototype being modelled. In most instances, modellers will use the prototypical metals to construct the relevant parts. One area where this is not practical is a model that requires chromium-plated parts. When building scale-model cars, motor cycles and their engines, electro-plating is often essential. A typical example is the exhaust shown in Photo 9.5. Whilst chromium plating is not impossible in the home workshop, it requires the part first to be copper plated

CHAPTER 9: METAL COLOURING

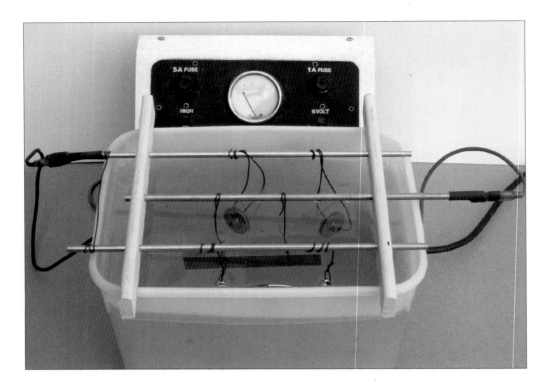

Photo 9.6: Anodising some components in a sulphuric-acid bath made from a two-litre plastic container

using sulphuric acid and then nickel plated before the final layer of chromium is added. And the process includes cleaning (any surface contamination will spoil the resulting plating), rinsing and polishing at each stage. This requires containers to hold the chemicals, a means of suspending the items to be plated, together with pieces of each of the sacrificial metals, in the various solutions. It also involves passing an electric current through them from a variable-current power supply, as well as thermostatic heaters for the chemical solutions. In addition, cleaning and rinsing containers are needed. The time taken to plate the items depends on the thickness of plating required, the plating metal being deposited and the size of the item being plated. In practical terms, while several companies offer complete plating kits, including all the necessary tanks, these limit the maximum size of part that can be plated. Perfect results are thus much more likely to be achieved if parts are professionally plated.

ELECTRO-LESS PLATING

Also known as chemical plating, the technique employs a suitable solution of a chemical containing the metal to be deposited and a reducing agent. Plating can take place without any electric current; the component to be plated acting as a catalyst. The resultant plating is hard and evenly distributed. Using this technique, nickel can be deposited onto ferrous metals and aluminium, as well as on copper and copper alloys when some iron or aluminium wire is needed as a catalyst. And a thin layer of tin can be plated over all ferrous metals, copper and copper alloys. It is a pity that the constituent chemicals required and the by-products of this process make it unsuitable for home-workshop use.

ANODISING

Anodising is a way of treating metals mainly associated with aluminium, although magnesium, titanium and zinc can also be anodised. When exposed to the atmosphere, aluminium forms a passive-oxide layer which gives quite a reasonable degree of protection against corrosion. However, any aluminium alloys with a magnesium content are very prone to atmospheric corrosion and benefit from being anodised.

As well as being protected by the process of anodising, aluminium and its alloys may also be provided with a decorative coloured finish. The purer the aluminium is the better the results. Silicon or manganese in the alloy will slow the anodising process, and some alloys and castings tend to give poor results.

The work piece is connected as the anode in an acidic solution and a current passed through it to form an oxide coating. In the case of aluminium, the process forms a layer of aluminium oxide, Al_2O_3, or corundum, which is very hard, quite inert and is able to absorb coloured dyes. It also provides electrical insulation. Anodising is a process widely used on cameras, torches

METAL FINISHING TECHNIQUES 91

Figure 9.7: A seven-cylinder model aero engine with colour-anodised rocker covers and push-rod housings

and MP3 players and is well suited to some modelling applications. Unfortunately anodising is prone to cracking if its temperature exceeds 80°C.

The protective-layer thickness depends on the voltage used during anodising, while the pore size varies with the concentration of the acid, its temperature and the electric-current flow. As the layer is only going to be a few tens of microns (a little over 0.0001") thick, it will thus only have a slight impact on the dimensions of the item being anodised. More importantly, it will slightly affect the size of any holes in the metal for rods, pins or bolts. The holes for the first two can be drilled a fraction over size while threaded holes need to have the appropriate size of tap re-run through them. The finish of the surface of the item being anodised affects its final appearance. As a result, any defects like corrosion or polishing burns are likely to be enhanced rather than being concealed.

Anodising is undertaken at 20° - 25°C in a 10% - 15% solution of sulphuric acid. It requires the components to be de-greased and thoroughly cleaned before being connected as anodes using aluminium wire with lead sheet for the cathode. With electric power at 12 volts and 10 - 15 amps per square decimetre (foot) it takes around ten minutes to produce a layer 0.0025mm (0.0001") thick. Agitation by means of a fish-tank bubble compressor will improve the resultant anodising. The component should then be rinsed in clean water. And the resulting layer can be sealed by immersing the item in boiling de-ionised water for twenty minutes.

COLOURING ANODISED ALUMINIUM

The result of anodising almost any aluminium alloy is to provide a silver-coloured protective surface. However alloys that contain 5% or more of copper, manganese, magnesium or silicon will colour when anodised; orange/yellow with copper, brown with manganese and blue/grey with magnesium or silicon. For many other colours, organic dyes are suitable. They should be well dissolved in water at 50°C in a tank with bubble agitation and the part suspended for between five and fifteen minutes, depending on the depth of colour desired. Photo 9.7 shows a classic example of an aero engine which has rocker and push-rod covers that have been anodised bright red but appear black in the illustration.

APPLYING PATINAS

A patina is commonly an oxide or carbonate that forms on the surface of metal exposed to the elements. A patina also refers to changes that occur in the surface texture

Photo 9.8: The patination is visible on many of the parts of this fine static engine

and colour of metal as a result of normal use, for example in old bronze coins.

Constructors sometimes want to add a patina on parts of a model to avoid a shiny, fresh-out-of-the-workshop appearance. Some patinas also protect the surface of the metal against further corrosion.

Producing most colours involves a range of unpleasant chemicals and hot or cold processes. However, copper and its alloys, brass and bronze are the easy to age with a proprietary antiquing fluid. Many colours can be produced; the popular ones for engineering models are black, brown and green. The process of bluing to provide a black colour on ferrous metal has already been described on page 87. For the other two colours, cold chemicals can be used. The green patina or verdigris that forms on bronze and copper is copper carbonate.

To produce a rich brown patina on copper or its alloys, mix 5g (0.2oz) of ferric chloride and 2.5g (0.1oz) of ferric nitrate with half a litre (one pint) of distilled water and then evenly apply the solution to the metal surface with a brush or sponge and leave it to dry. When a pale-brown colour appears, rinse the metal well with cool water and dry with damp newspaper. Leave the coating to harden over night and then repeat if a darker shade is wanted. Once the required colour is reached and the piece has dried over night, finish it with wax to darken and set the colour of the resulting patina.

Alternatively, use a brush or cotton wool to apply a proprietary solution such as Liberon Antiquing Fluid directly on the item. Once the desired colour is achieved, immediately rinse with clean water and dry. Otherwise, dilute the solution with ten parts water and immerse the items for a maximum of two minutes until the desired brown colour is reached. Then apply a water-displacing oil, such as Liberon Jade Oil, to fix the colour.

Proprietary green patina solutions, such as those from Jax and Sculpt Nouveau, may be brushed onto the part or it may be dipped in the cold solution. The result is a pleasing pale green that matches the verdigris on many full-size projects.

Gilding wax consists of fine metal particles coated with a binder of wax and acrylic resin. The combination of patinating bases and gilding waxes creates a rich range of different patinas. The wax is suitable for applying on metal as well as to wood.

10: PAINTING

PAINT FINISHES

The vast majority of models and other items built in the home workshop will, to some extent, require a paint finish, either to match a prototype or to protect the metal surface or both. Many items require a series of coats of different paints before the final finish is applied and many may need more than one layer of the final coat to achieve the best results. And it is essential to let paint dry thoroughly before handling, adding lining or lettering or dealing with any surface problems.

The aim of this chapter is to examine the wide range of different potentially suitable paints and varnishes and indicate how they are applied to obtain a fine finish. No attempt is made to try and explain how to ensure authenticity for scale decoration of a complex model such as a traction engine or steam locomotive.

Even before thinking about painting a model, bear in mind that any lack of smoothness or imperfections in the surface of the metalwork will not be disguised by layers of paint. Thus a perfect bare-metal finish and one that is also totally clean should always be the first aim.

Remember that practice is crucial and be prepared to experiment with different paints, tools and metals to get the optimum result. Always try to paint in a dirt- and dust-free space that is not over humid and one where

Photo 10.1: Bongo, the locomotive that spawned the book 'How (not) to paint a locomotive' by Christopher Vine

Photo 10.2: A range of flat, pointed and mop brushes of different shapes and sizes

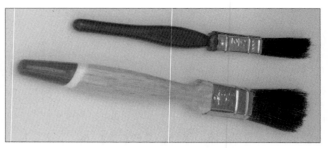

Photo 10.3: Two small DIY brushes that measure 9mm (0.375") and 12.5mm (0.5")

the metal is not too cold. Painting is best carried out at a temperature comfortable for the painter. And if spraying paint, some sort of spraying booth is virtually essential.

The final coat or coats of paint are likely to be of enamel or cellulose (also referred to as lacquer), with acrylic, epoxy and polyurethane all coming a distant second. Many paints are single-part air-drying that rely on solvent evaporation. Organic solvent-based paints tend to adhere better than waterborne paints while the latter are better in environmental and health and safety terms. As an alternative, a two-part acrylic, epoxy or polyurethane may be used though these tend to have their own drawbacks. Drying is caused by a chemical reaction between the two constituent parts. Both single and two-part paints must be thoroughly mixed before they are applied.

The glossiness of paint or varnish depends on the amount of light reflected by its surface when it is dry. It may have a gloss, semi-gloss, satin or matt finish. Matt paints have non-reflective properties to give a flat finish that does help to hide surface imperfections. Satin or eggshell finishes gives a slight sheen, resembling the shell of an egg. They provide a harder, more durable surface with better stain resistance than matt paint. Semi-gloss paints are durable, relatively easy to clean and resist staining better than a satin finish. Gloss paint is the norm for most of the parts of engineering models and provides a tough, hard-wearing, stain-resistant finish that is easy to clean.

Unfortunately, some paint types are not compatible with others resulting, in the worst cases, in bubbling of the paint that ruins the finish. Sometimes, poorly applied coats of paint need to be removed. If done immediately, paint solvent will often do the job. Otherwise emery paper or a suitable paint stripper is needed. Unfortunately, most of the latter require care in handling as well as in disposal of the stripped paint.

It is worth understanding that, when painting a scale model, even a paint chip from the prototype will not provide exactly the correct colour. As the distance from any full-size prototype increases, its colour becomes less bright. To demonstrate this, compare the colour of trees on the distant horizon with the colour of the same type of trees nearby. The green appears paler on the distant trees. When building a Gauge One scale locomotive, looking at it at a distance of one metre or yard is like observing the original at a distance of thirty-two metres or yards. And there will be a slight colour change; the smaller the model's scale the more noticeable the effect.

Some paints produce poisonous fumes and are not recommended for home-workshop use. And most paints, varnishes, thinners and paint strippers are poisonous, many are inflammable and all cause damage if spilled. So it is essential that they are kept out of children's reach.

THE EQUIPMENT NEEDED

BRUSHES

A good paint brush is worth its weight in gold and artists' water-colour brushes fulfil most requirements. For the finest work the best sable brushes are not cheap to buy. However, alternative animal hair, such as hog, or synthetic-bristle brushes like Dalon or even a mixture of animal hair and synthetic bristles offer excellent alternatives at much more affordable prices. There is also the need to have the right size of paint brush for the job and that depends on the task in hand. Artist brushes come in numbered sizes from 0000, the smallest, to 24; round with pointed ends or flat with chisel ends. For larger brushes, there is at least one UK artist's range that offers flat brushes, originally designed for sign writers, in imperial sizes from one eighth to one inch in one eighth steps (3mm to 25mm in 3mm steps). DIY paint brushes are flat and are still in imperial units. Useful sizes start at 0.25" (6mm), then 0.375" (9mm), 0.5" (12.5mm), 1" (25mm) and larger. In principle, try and use as large a brush as is practical since it will carry more paint and, in all probability, have softer bristles. However, using too large a brush for fine work is equally a mistake.

A stippling brush is specialised and is used to provide soft edges. One can be made by cutting most of the hair off the end of a brush to leave around 6mm (0.25"). Brushes specially designed for painting thick or thin lines have very long thin heads that hold plenty of colour.

SPRAYING EQUIPMENT

Undoubtedly spraying produces better results than brushing. But it requires better preparation in terms of masking areas not to be painted and superior facilities – at least a spraying box and a mask – and unless using aerosol spray paints, more equipment and cleaning at the end of each painting session. For the vast majority of smaller engineering models, a good-size air brush will suffice but for larger scale models a small spray gun is needed. Both types of spraying equipment require a compressor with an air reservoir to power them.

Both airbrushes and spray guns work by passing a stream of pressurised air through a venturi that generates suction to draw paint from their reservoirs. The rapidly moving air atomises the paint into minute droplets as it passes through the paint-metering assembly. The paint then sprays onto the metal or other surface. The paint flow is user-controllable by an adjustable trigger that moves a fine tapered needle in the spray nozzle. And the spray pattern can also be altered.

The two normal ways of feeding paint are by gravity feed from a reservoir sitting on top of the airbrush or spray gun, or by suction from a reservoir mounted on the side or underneath; side or bottom feed. Gravity has the advantage that it puts the paint into the mixing chamber, requiring less air pressure to suck the paint and providing the finest paint atomisation. Side- and bottom-feed tools provide the user with a clear view over the top. Bottom-fed units can usually accommodate a larger reservoir. Reservoirs often incorporate a built-in filter to remove any contamination.

Airbrushes

An airbrush comprises a cylindrical tube with a connection at one end to an air compressor and an adjustable spray

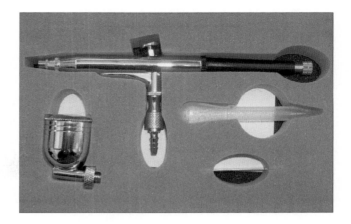

Photo 10.4: An air brush suitable for painting smaller models

head at the other, together with a refillable paint reservoir. Airbrushes can be grouped by their features. How is the paint flow triggered, how is the paint fed to the airbrush and how are the paint and air mixed? The first two characteristics define whether the airbrush is single or dual action. On a single-acting airbrush, pushing down a trigger on its top sprays out a paint/air mixture and releasing it stops the flow. An adjustment is provided to vary the paint to air ratio. On a double acting unit, pressing the trigger provides air and pulling it back then also releases paint.

Airbrushes are designed for artistic work but are also ideal for spraying the parts of all but the largest models. Their paint reservoirs typically hold around 20ml (1oz) of paint. For larger models, a spray gun will cover the increased areas involved in a reasonable time.

When looking for a quality airbrush there are several points that should be taken into consideration:

1. The controls should include a smooth, preferably double-action trigger and should be able to give precise control of the paint flow.
2. All the internal passages should be solvent-proof and the needle housing needs to be made from a some form of PTFE such as Teflon.
3. The airbrush should be able to spray the range of desired paints such as cellulose, enamels and water-based paints.
4. It should be comfortable to hold, light and well balanced, even with the air-supply pipe connected.
5. Disassembly and cleaning of the airbrush should be straightforward.

Two airbrushes with fine reputations for painting models are the DeVilbiss Sprite and the Badger 200.

Photo 10.5: The Osprey spray gun is large enough to paint the parts of the biggest models

Spray guns

The majority of spray guns have die-cast pistol-shaped bodies and have three controls to allow the optimum performance to be achieved for any chosen application. These controls adjust the rate of paint flow, the size and shape of the spray pattern and the pressure of the air passing through the gun. There is also a trigger to turn the spraying process on and off. And for a spray gun, a gravity-fed paint reservoir makes it easier to clean and to change to different coloured paint.

Spray guns require a significantly greater volume of air than airbrushes but even small ones are suitable for painting most component parts of 7¼" locomotives and quarter-scale traction engines. Thus care is needed not to choose too large a spray gun and to select a matching air compressor that delivers a sufficient volume of air. Inevitably, choosing the right gun for the tasks likely to be undertaken is not straightforward. There are several different types and classes.

HVLP, or high-volume/low-pressure guns employ a relatively high volume of air delivered at a moderately low pressure to atomise paint into a soft, low-velocity pattern of particles. HVLP guns operate at an air pressure of typically up to 2 bar (28 psi) with only around half that pressure at the nozzle. And a small one will only consume around 140 litres/minute (5 cfm).

'Normal' spray guns operate at significantly higher pressures and eject the paint mist from their nozzles at a much higher velocity. This does make achieving a fine

Photo 10.6: A compact 60 l/min (2 cfm) 3 bar (45 psi) electrically powered air compressor

finish on small components harder and will also waste more paint.

Regardless of the gun chosen, a selection of different nozzles is desirable to achieve the required spray pattern and to minimise paint wastage.

Compressors

Reciprocating air compressors provide atmospheric air to a given pressure and rate of flow. The type of unit needed will depend on the kind of spraying equipment to be supplied with air. Most airbrushes or spray guns can be purchased complete with air hose, connectors and air compressor. When selecting a compressor for use with an existing airbrush or spray gun, check the average air consumption and pressure required and choose an appropriate unit that can provide at least the volume and pressure required but preferably rather more. A good-size reservoir will be needed to ensure that the gun is never starved of air. And there is a preference for an oil-less compressor to prevent oil contamination of the air.

The air pressure needed varies depending on the size and design of the spraying equipment; typically in the range 1.5 – 4bar (20 – 60psi) for the smaller units needed to paint limited surface areas of the models and tools that are the output of the average model engineer's workshop. It is essential to ensure that a compressor can deliver sufficient air to power the airbrush or spray gun. The compressor in Photo 10.6 will power an air brush whereas the spray gun shown in Photo 10.5 needs as much as 225 – 350 litres/minute (8 – 12 cfm). A typical DIY compressor will be fitted with a 1.5kW (2hp) electric motor, a 25 or 50 litre (1 to 2 cubic feet) air

METAL FINISHING TECHNIQUES 97

Photo 10.7: Three different air compressors, each with a separate air reservoir, that will deliver 125, 200 and 280 litres/minute (4.5, 7 and 10 cfm)

Photo 10.8: A cardboard-box 'spray booth' is suitable for all but the largest components

reservoir or receiver and occupy a space of around one fifth of a cubic metre (6 cubic feet).

Two enemies when spraying paint are any oil drops from the compressor and water drops from the atmosphere. Both can be eliminated by fitting suitable filters in the air line. An oil-free compressor eliminates the problems of air contaminated with tiny drops of the lubricant.

Spraying booth
A small spraying booth, made from a large cardboard box, is suitable for the components of all but the biggest models. Photo 10.8 shows a box, open at the front, that includes a turntable so that the items to be sprayed can be rotated to give access to all sides except the base on which they rest. The wooden jig allows the flywheel to be rotated during spraying; its hub and rim having first been masked. For larger components, a polythene-sheet-lined, wooden-framed booth may be built or a section of the workshop screened off. A decent mask is essential to avoid inhaling paint spray as are overalls and thin gloves to keep paint off the skin.

Miscellaneous items
In addition to the correct equipment and clothing, spray painting will require some or all of the following items: thinners (the type depending on the paint used), clean cloths, tack cloths, cotton wool buds, masking tape, brown paper and frisk film (see Page 108).

LINING TOOLS
For those who plan to undertake a lot of lining of their models, investing in some form of lining tool is crucial. There are several ways of producing lines but the most effective is to use a professional tool with a set of heads to produce lines of varying widths.

The Beugler lining and striping tool is ideal for model-engineering projects. It can produce uniform painted lines in widths of 0.4 mm, 0.8 mm, 1.3 mm, 1.7 mm, 2.5 mm and 3.2 mm (one sixty-fourth to one eighth of an inch in sixty-fourth steps), and up to 12.7 mm (0.5").

Photo 10.9: A pair of Beugler lining tools with interchangeable heads, for lines of different widths, and adjustable guides

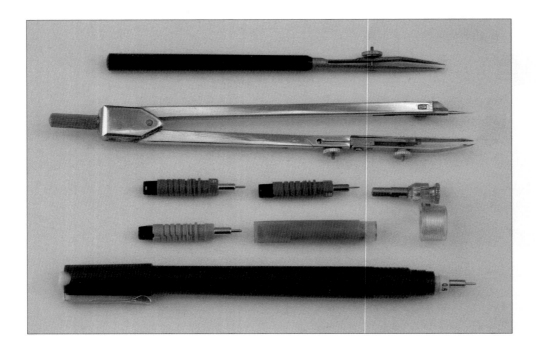

Photo 10.10: A bow pen may be used for straight lines, the matching compasses for circles. Below are different size nibs, a spare reservoir and compass fitting for the Staedtler technical pen shown at the bottom

Interchange of wheel heads makes altering the line width an easy task. As the wheel rotates it picks up paint in its serrations. The tool employs a 'syringe-like' plunger to feed paint to the wheel. It can be used in any position or orientation. A range of guides is provided with the tool so that an edge or template can be followed to make straight or curved lines. However, such a precision tool is not an inexpensive choice.

Bob Moore's Master Lining Pen comprises a pen handle with a ball-jointed pen head incorporating a small paint reservoir. There are three head sizes; fine that produces a line about 0.2mm (0.008") wide, standard 0.34mm (0.012") wide and standard plus 0.5mm (0.020") wide. There are also two guides to help in following any suitable edges as well as a ground-glass syringe to aid cleaning the pen.

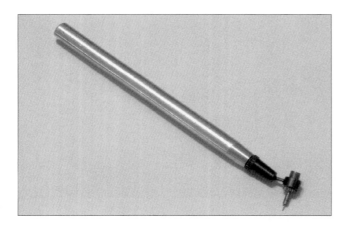

Photo 10.11: Over 10,000 Bob Moore lining pens have been sold worldwide

Alternative low-cost solutions are to use a draughtsman's technical-drawing pen with a set of interchangeable heads to give different line widths or employ a bow pen for straight lines and a matching pair of compasses for circular ones. Thinned paint rather than ink should be carefully fed into the pen reservoir or bow.

SURFACE PREPARATION

Good surface preparation is the key to obtaining a first-class paint finish on any model. It will need to be disassembled into its constituent parts before painting can start. A set of well-made component parts is not enough; the parts must also be thoroughly cleaned with acetone to remove any traces of oil or grease. And clean also means without any finger marks which deposit natural oils and greases from the skin onto the metal. Do not use washing-up liquid for cleaning as it leaves a film that is incompatible with enamel paint.

A clean and dry surface is the first and most important pre-requisite to starting painting. The section on page 7 that deals with cleaning and pickling provides some helpful advice. For non-ferrous metals, first scrub with detergent and water, careful drying with a lint-free cloth. Then wipe over the whole component with a suitable solvent; white spirit or cellulose thinners. For ferrous metals an alcohol-based liquid, such as methylated spirits works well, though care is needed as it is both volatile and highly flammable. And ferrous metal, in particular, requires scrupulous cleaning to remove any products of the rusting process. If the metal is to be varnished,

METAL FINISHING TECHNIQUES   99

Photo 10.12: A nicely constructed 0-4-2 tank engine ready to be stripped down for painting

this can now be undertaken but if it is to be painted, it will first need to be coated with a suitable primer.

Metal surfaces that have been polished will require a final-cleaning process before they are varnished. In the case of brass, probably the most common polished metal to be protected from tarnishing by varnish, an extra 'dry polishing' stage is needed after cleaning and polishing to remove any deposits left by the polish. Feldspar, chalk or flour may be employed for this last stage using a soft cloth.

The purpose of a tack cloth is to remove any final remaining particles of dust or dirt from any work piece when wiped over the surface of the cleaned metal. Tack cloths are sufficiently sticky to pick up dust but do not leave any residue on the object being cleaned. Tack cloths are widely used both in woodwork where sawdust is a real problem and by re-sprayers of cars. They are also applicable to metal models, particularly those with large areas to be painted.

RUSTY METAL

Rusty iron or steel is, hopefully, rarely found in a well-managed home workshop. However, models and tools that are being restored (or have experienced a long gap mid-construction) will require any existing rust to be removed before proceeding further with the cleaning process. Rust-removing dips, gels and liquids that can be painted onto the rusty surfaces are made by a number of companies. Granville Rust Cure, Hammerite Kurust, Dinitrol RC900, Rust-Oleum Rust Reformer and Rust Stripper are good examples of proprietary brands. Car accessory shops are likely to carry a range of products and the manufacturer's instructions should carefully be followed. Multiple applications may be required to remove all rust. After removal, bare metal should be oiled, greased or protected with several layers of an appropriate paint.

Do not start painting without properly preparing the surface, since a short-cut will compromise both the appearance and durability of the finished paint. Rust begins to form almost immediately when ferrous metal is exposed just to moisture in the air. Even if the metal is new without visible rust, surface preparation is still necessary and any traces of oil must be removed before starting painting. While it may prove to be impossible to eliminate all traces of old rust, as much as possible should be removed before applying a primer. A quality, acrylic

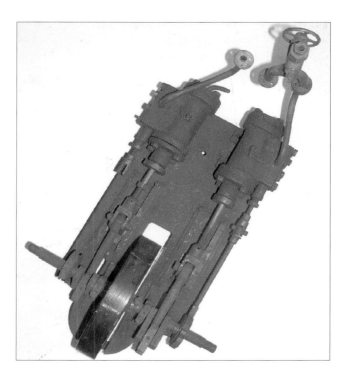

Photo 10.13: Left unpainted, the steel parts of this static steam engine have rusted all over.

Photo 10.14: Apart from the boiler that still awaits cladding, and the wheels, the rest of the model has been carefully primed

latex or oil-based corrosion-inhibiting primer will prevent rust reappearing. It is worth taking the extra time with surface preparation and then applying the chosen primer immediately afterwards.

FILLERS AND PRIMERS

One of the major problems with the use of either fillers or primers is their ability to mask some details in the surface of the metal; rivets for example. Thus a thin coat of any primer is desirable despite manufacturers usually advising either thick coats or several thinner ones.

Fillers

There are several fillers that can be used on most metals and many of them have been formulated for use in the automobile body-repair business. They are perfect for covering blemishes in castings and may be helpful in making repairs when machining errors are made. Of course some errors can readily be filled with silver solder.

Three common trade names are Isopon, Chemical Metal and Milliput; the first two are polyester fillers, the last an epoxy putty. Both of the fillers are two-part grey-coloured pastes that need to be mixed together in the appropriate quantities. They harden quite quickly giving off a distinctive odour. Isopon P38 can be sanded and painted after it has set in just twenty minutes. Plastic Padding Chemical Metal is resistant to water, oil and petrol, sets in the same amount of time and can be sanded, machined, drilled, tapped or polished. Milliput comes in a range of colours, of which silver/grey is the most appropriate. Mix equal parts of the two sticks for five minutes and continue for a further minute after the two different-coloured parts merge into a single uniform colour. It sets in three to four hours depending on the temperature and is fully cured in twice that time when it can be machined, drilled, tapped, turned, filed, sawn, sanded or painted. All three materials should be applied to clean, grease and dirt-free surfaces, preferably abraded to increase adhesion. Once completely set, they should be sanded down until, by running a bare finger over the repair, the metal surface is entirely smooth. Once set, fillers are difficult to remove from tools so clean them before the filler has completely cured.

Primer fillers

These fillers have a higher solids content (usually talc) which provides the filling properties. They allow filling of small scratches and defects left by any previous preparation, but not large ones. There is a choice between solvent-evaporation and two-pack primer fillers.

Solvent-based products are largely cellulose or acrylic and their main advantage is their fast drying time. Their shortcomings are low filling power, adverse reactions with underlying coats and sinkage if not dried properly. They are available in aerosols or tins, and the latter need to be thinned, usually 50/50, with the appropriate thinners. Application is normally two or three coats with adequate drying time between coats to ensure the solvent has evaporated. Avoid heavy coats and try to use a greater number of thin coats.

Two-pack primer fillers include a larger solids content, which, together with their resin base, gives far better filling power and fewer problems with sinkage. They are mixed just before use, carefully following the manufacturer's instructions for quantities. They have a limited usable life that depends on the temperature as well as the type and amount of activator used. Up to three coats may be needed, depending on the surface being filled, with time required for each to chemically set. However, they have some excellent properties to compensate for their lengthy drying times. Most two-pack primer fillers are thick and heavy even when thinned so that a relatively large nozzle may be required for spraying. All primers and primer fillers must be sanded down when dry.

Primers

Bare metal does not take paint well so that a primer of some sort is an essential pre-requisite as a preparatory coat. The primer is a foundation layer that is applied before adding any further coats of paint to components. Typical examples include red oxide, red lead, zinc rich and zinc phosphate. Priming improves the adhesion of paint to the metal, increases paint durability and gives extra protection to the material being painted.

Priming is especially important when painting iron or steel and it is essential to use the right type of primer. Not only will it help prevent corrosion and rusting, but it will also assist subsequent layers of paint to adhere to the metal. However, check that the primer is compatible with any proprietary rust treatment that has been applied.

Other metals used to make models require priming before they are painted. These include the various forms of aluminium, brass, bronze, copper and zinc as well as other less common metals that find occasional use and, for one reason or another, require a paint finish.

Etching primers

It is essential to use an etching primer, which will etch itself into the surface of the material, on non-ferrous metals; particularly brass. Most are acid-based and are supplied as a primer base plus an acid activator that is added before use. Some single-component etch primers are available in aerosol form but are less effective than acid-based ones.

Choose an etching primer to match the metal to be primed and whichever type is used, follow the instructions provided and keep the mixture in a plastic rather than a metal container. Plain etch primers without any filling properties should be applied as a single coat over

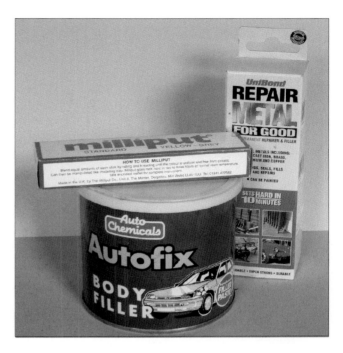

Photo 10.15: Three different brands of chemical filler; two polyester-based and one an epoxy putty

the bare metal. The etch coat must be quite dry before continuing to add further primer filler or paint coats.

Etch primer fillers provide modest filling capabilities and two or three coats should be applied with ten to twenty minutes between coats so the solvent can evaporate.

Do not, under any circumstances, use any thinners with etch primers or etch primer fillers apart from the ones that are recommended by the manufacturer.

PAINT CHOICES

Many factors determine the best type of paint for any particular task. These include the colour required, the metal to be painted, its type of surface and condition and the age and type of any paint (usually primer) that has previously been applied. As health and safety regulation have become more dominant, attempts to get rid of volatile solvents have progressed significantly. Unfortunately, water-based paints are far from ideal for painting models, particularly those with ferrous-metal component parts.

Paints are usually made from two main constituent parts; the pigment which give the paint its colour and the binder that provide adhesion, holds the pigment together and affects properties like durability, glossiness, hardness and flexibility. In addition, two other ingredients may be

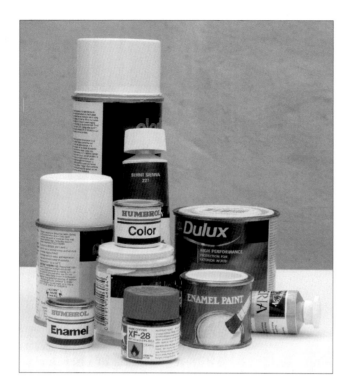

Photo 10.16: Paints are sold in tubes, in jars, in cans and in aerosols

found; solvents that decrease the viscosity of the paint and additives that may alter characteristics such as surface tension, flow properties or appearance. As illustrated in Photo 10.16, paint may be bought in tins, tinlets, jars, aerosols and tubes.

Historically, paints used by modellers have been either cellulose or enamels with a petroleum-based solvent. Acrylics were taken to be water-based (aqueous acrylics popular with artists) and many still are. But a number of acrylics now use a chemical solvent. Acrylic resin is a colourless, transparent thermoplastic used to make both lacquers and enamels. These have the great advantage of not needing to be buffed to obtain a shine.

Today there are two main classes of paint likely to be used to finish engineering models; solvent-based paints and chemical setting two- and one-part paints. Solvent-based paints tend to separate during storage, the heavier pigment settling to the bottom. They therefore require thorough mixing before they are used.

Chemical setting two-part paints, such as epoxies, also need to have the two parts well blended together. They have a short pot life once they are mixed but they provide hard, tough, glossy finishes. Normally they do not involve the use of any solvents. There are also a few one-part paints that also chemically cure on exposure to air.

ENAMEL

Enamel paints are undoubtedly the most popular when the time comes to decorate engineering models. They adhere well to primed surfaces and to some bare metal. They are relatively easy to apply though they do emit a distinctive odour. They dry to a hard and durable finish; typically in 8 to 24 hours but some much faster.

Enamels are available in an almost unlimited range of colours; UK companies like Humbrol and Precision Paints specialise in making enamel paints for use by modellers. These can be purchased in aerosols or in relatively small tins; 100ml (3.4fl oz) or smaller. Enamels are normally glossy but matt finishes are popular for some models. And the range of colours is impressive, many having been carefully specified to match the original colours used by full-size prototypes; particularly steam locomotives. Machine-tool companies like Myford sell touch-up paints for their machinery that match the original colour, either in tins or aerosols, for those wishing to undertake repair or refurbishment.

Enamels dry by evaporating a white-spirit-based solvent in the paint and typically require up to twenty-four hours to dry and harden completely. Enamels are readily thinned for spraying though their lengthy drying times make this a tedious process. Brushes must be cleaned with white spirit or turpentine. Water-based enamels are not recommended for painting metal. Even though they dry more quickly and are more environmentally friendly, they do not adhere as well nor are they as durable.

CELLULOSE

Cellulose paints were, for almost all of the last century, the choice of paint for the automobile industry. They provide a fast-drying glossy finish and are readily thinned but at the cost of a strong, highly-inflammable odour when being applied. Health and safety regulations mean that they are disappearing from use and availability tends to be limited to repairs and touch-up of older cars.

A wide range of colours may be obtained from a few suppliers but the majority can only provide colours used on existing cars and either in 500ml tins or in aerosol cans. Brushing cellulose is not easy because of the speed of drying so spraying is much the preferred method of application. And if a tin is to be used, it will need to be diluted with acetone-based thinners before being sprayed. Unfortunately cellulose paints will attack any enamel base coat as well as most plastics.

Despite these issues, cellulose is still popular because it dries so rapidly that several coats can be applied in a

METAL FINISHING TECHNIQUES   107

Photo 10.21: A holding jig of dowel and scrap ply on a turntable for spraying a masked flywheel

where the solvent can evaporate; also one that is not too cold or too humid; the latter can cause the paint to bloom.

HOLDING THE WORK
Whether planning to use a brush or to spray, serious consideration is needed to decide how to hold each part while it is painted. Depending on the size and shape of the item, it may be mounted on a length of wire or dowel, screwed to a long bolt or suspended by string or wire. It may also be placed on a face that does not need painting or has already been masked. In each case, the way of providing support for the item during and after painting also needs careful thought. A turntable can be very useful, particularly when spraying items, allowing them to be rotated so that all sides can be accessed.

MIXING
While many paints can be used straight from the can, they generally require thorough stirring to ensure any pigment that has settle to the bottom of the tin is well mixed. Failure to completely mix paints before use can result in an inaccurate colour, an incorrect surface appearance (matt paint can end up shiny and vice-versa) and even poor adhesion. Some paints, especially cellulose, require thinning before application and particularly before spraying. Equally important is complete mixing of two-part paints in throw-away containers, such as cleaned glass or plastic food or liquid containers. But do check that the paint does not melt the plastic.

Use a pillar drill to stir up paint, particular any that has been stored for a long time, to significantly improve the mixing process. This is illustrated in Photo 10.22.

Photo 10.22: Mixing using a bent-wire rod in a pillar drill requires a slow rotation speed

Before starting work, think about the thickness of the paint and the number of coats to be applied. Thicker paint provides better coverage but does tend to mask any fine detail. On the other hand, thin paint does not cover so well, meaning there is a need for multiple coats with more risk of dust or insects settling on the wet surface and damaging the finish.

BRUSHING
The excellence of a brushed finish will depend on the brush quality, having the correct viscosity paint and, to an extent, type of paint and the skill of the painter. Too small a brush limits the amount of paint it can carry to whatever is being painted and almost always results in an inferior result. Most paints should be well brushed into the surface but a few need only a minimum of brush strokes. Always check the maker's recommendations given on the container. Successful painting demands a well-lit and ventilated, dust-free working area. Brush marks can be minimised by painting alternate layers at ninety degrees to each other. It is essential to maintain a wet edge to the paint if covering a large surface area to ensure a smooth finish.

MASKING
Any model where its components are to be sprayed may require parts of these items to be masked. Brown wrapping paper held in place with masking tape is one

Photo 10.23: Much masking will be needed before painting this locomotive. Note the engine-turned surround to the firebox door

solution; with newspaper the printing ink may mark the masked surfaces. Masking tape comes in many different widths and the thinner ones are quite good at forming curved edges. When removing masking tape, either take it off as soon as the paint has been applied or wait until it is completely dry. Always pull the tape back on itself rather than pulling it vertically up from the surface to avoid removing any paint under the tape.

The other alternative is to use frisk masking film; a low-tack, self-adhesive transparent material. First cut the film to size using a sharp scalpel on a suitable cutting surface, preferably with a paper template trimmed to the exact size needed. Then apply the frisk film by removing the backing paper and laying the film on the surface to be masked. Start in the middle of the film and rub it down, working out any air bubbles towards the edges of the film. This task can be eased by using the edge of a credit card to push out any air bubbles. In addition, to help mask small and complex-shaped components, masking fluid, a latex-based liquid, can be painted on prior to spraying and then readily peeled off.

SPRAYING

There is little doubt that spraying produces a better result than a brush, particularly on smooth surfaces. For the small amount of paint involved in finishing many smaller models, aerosol cans are a worthwhile option. These cans contain an agitating ball that must be shaken for a couple of minutes to ensure that the paint is fully mixed. After spraying, the can should be turned upside down and squirted for a couple of seconds to clear any paint from the nozzle. The can will then be ready for further use, avoiding it getting blocked by dry paint.

Compared with brushing, spraying needs better facilities; a spraying booth or box, face mask and protective clothes as well as an air brush or spray gun and a compressor, unless using an aerosol. More cleaning is involved after each painting session and spraying involves masking any parts that require a different colour paint or are to be left bare. For almost all smaller models, an air brush rather than a spray gun will be preferable; the latter only really being applicable to larger scale models.

Most paints are supplied ready for brush application and must be diluted with thinners before spraying to enable the paint atomise. The type of thinners can usually be determined from the manufacturer's information on the paint container or the type that is recommended for brush cleaning. As a starting point, try a carefully measured fifty/fifty mix of paint and thinners and mix the two in a separate container to ensure the correct thickness. Then pour, and preferably strain with a paper or nylon filter,

METAL FINISHING TECHNIQUES ■ 109

Photo 10.24: Well-sprayed paintwork can give a wonderful shine. Reflections in the paintwork on the side of this scale steam lorry are clearly visible

into the airbrush or spray-gun reservoir. A surplus of thinners may cause drips, runs, sagging or even pinprick bubbles as excess thinners escapes and will require additional coats to achieve the desired cover. Even if problems do not occur, more coats will be needed to achieve the desired coverage and finish. Too little thinners will result in the nozzle blocking (as will dirt in the paint or dust in the reservoir) or in an uneven paint surface.

A secret of spraying is to apply several thin coats and allow them to dry between coats. A few thick coats will almost inevitably result in unsightly runs that ruin the finish. Aim to obtain a solid cover of paint in around three coats.

It is important to set up equipment to give the desired spray pattern and correctly atomised paint. The normal aim is an elliptical spray pattern with a completely wet centre part and an outer edge of tiny droplets of paint. Start by setting the paint-control knob wide open and a similar completely open position for the internal air flow. This allows fully atomised paint delivery, with the correct air/paint ratio, to be distributed in the chosen spray pattern. Also set the right air pressure and volume in accordance with the manufacturer's suggestions.

Connect the airbrush or spray gun to the air supply and fill the paint reservoir. Then check the spray pattern on a piece of scrap paper fixed in a vertical position. If the droplets of paint are too large, the air pressure is too low for proper atomisation and either needs to be increased or the rate of paint flow reduced. The latter solution also gives a smaller spray pattern.

Always hold the airbrush or spray gun perpendicular to the work surface. Try to maintain an even distance of not more than 200mm – 250mm (8" – 10") from the work at all times, constantly keeping the spray on the move. If the distance is too great, the paint will be too dry when it hits the component and the result will be a rough surface finish. Too close and the paint is likely to run. For larger components, a series of passes back and forth will be required to cover the whole item. And start spraying slightly to one side and above the component to ensure continuous cover with passes overlapping.

It is important to realise that a number of variables may affect the paint finish so that it is useful to note down

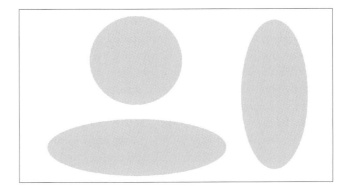

Drawing 10.25: Typical alternative circular and elliptical spray patterns that are easily selected

Photo 10.26: Soft-edge camouflage colours are needed on many armoured-fighting vehicles

the settings that give the best results. The key factors that influence performance are:

- Air pressure and air flow.
- Paint type and flow rate.
- Paint viscosity and amount of solvent added.
- Number of coats applied and their thickness.
- Time between application of coats.
- Spray pattern.
- Distance from nozzle to work piece.
- Speed of movement of the spray.

Cleaning an airbrush or spray gun requires copious amounts of the right solvent for the paint that has been sprayed; usually either acetone or white spirit. And the equipment will need to be stripped down to ensure that no paint remains within the internal components.

CAMOUFLAGE FINISHES

A rather different approach is needed if the original model has a camouflaged finish. Photo 10.26 illustrates clearly that the darker colour has been carefully applied over selected areas of the paler one. Where hard edges exist, either full masking of the paler colour before spraying the darker colour or the use of a brush are practical solutions. Where the edges are soft, a card mask correctly shaped and held a few centimetres (an inch or so) above the surface when spraying will produce soft edges. A short-haired stippling brush held vertically, applying small quantities of the paint with a dabbing motion, will result in similar blurred edges.

WORK BETWEEN COATS

Between each coat of paint, it is essential to lightly sand the surface with fine emery paper before applying the next coat. Care must be taken to avoid sanding right through a coat of paint, particularly if it is a primer. Special attention needs to be given when sanding sharp corners in metal work as the paint is easily pierced.

PAINT PROBLEMS

Runs and sagging

Any running of the paint or sags are almost invariably caused by the application of too thick a coat of paint; occasionally by thinning the paint too much. It is better to apply several thin coats rather than trying to apply a single thick one.

Orange peel or rippling is a problem found only with sprayed finishes. It is caused by the paint surface failing to flatten while drying. There are a number of causes including too low an air pressure and spraying at the wrong distance, paint that is too thick or just too thin a layer of paint.

Contamination

A quality paint job is readily spoiled by any type of contamination applied to the perfectly clean surface of a model. Much can be avoided by ensuring the paint is completely clean and filtering it if necessary. This is particularly so for paint in a tin that has already had part of its contents used. Hairs and bristle can fall out of even the best paint brushes but, if removed quickly, can be re-brushed over. On the other hand, dust is an eternal enemy that is less likely with fast-drying paints.

Oil and water contamination are both problems when painting with an airbrush or a spray gun. They can be eliminated with traps in the air line. The use of an oil-less compressor also helps. Colour bleed usually only occurs if a new layer of paint is applied before the one below has completely dried. It is also more likely when a pale colour is applied over a darker one.

Edges

Paint may be reluctant to stick to outside corners and also may accumulate to form too thick a layer on inside corners. The first problem is usually solved by initially applying a thin layer of paint to the edge, the second by care in the application of the paint.

Solvent problems

Too much solvent added to paint can cause pinholes or even blisters to form in the surface as the solvent evaporates. In addition, the solvent may attack the previous layer of paint causing it to wrinkle. In the worst case, the paint surface will harden but the solvent in the

METAL FINISHING TECHNIQUES ■ 111

Photo 10.27: Too thick a coat of paint has caused sagging on the side of the tender

paint below the surface will take a much longer time to evaporate, leaving the paint soft underneath.

Incompatible layers
Different types of paint are often incompatible and can cause the top coat to break up catastrophically. Furthermore, if the under coat is insufficiently dry, the solvent in the next coat may attack the lower layer, causing the paint to wrinkle.

Blemishes
Scratches, dimples, chips, dust and hairs can all, with care, be repaired. Some will require merely the lightest dab of paint from a fine brush, others will need a small area to be carefully rubbed down with wet and dry emery paper or a scratch pen and then re-coated. Once the final repair has completely hardened, it may then need to be cut back and polished using one of the many widely-available automotive products.

LINING
There are several ways in which lines can be produced. The best is to use a paint compatible with the layer on which the line is to be drawn, but not a fast-drying one. It must be heavily pigmented to completely cover the paint beneath, particularly when lining in white or yellow over a dark colour. An alternative is to use one of wide range of self-adhesive coloured plastic tapes that come in many widths and colours. For painted lines, there are several possible solutions. The first, easiest to use but also most expensive, is a commercial lining tool that will produce lines of the chosen thickness in place on the model.

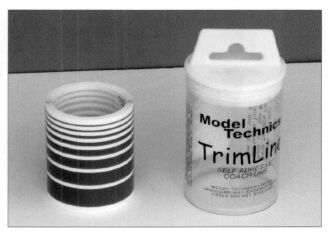

Photo 10.28: A roll of various thicknesses of 'TrimLine' plastic self-adhesive lining tape

To load the Beugler tool described on Page 97 and shown below, take off the wheel head and pull the plunger back to allow space for the paint. Use fresh paint that has a smooth consistency and be aware that getting good results is very dependent on the viscosity of the paint. Pour in the paint and replace the wheel head by pressing it firmly onto the paint barrel, taking care to align the wheel head with the guide bar. Hold the tool vertically with the wheel upwards and gently expel any air in the barrel by pressing the plunger upwards until a small bead of paint appears at the wheel. Wipe off any excess paint and hold the body of the tool near the horizontal. Using enough downward pressure for the wheel to roll, pull the tool to start it turning and to produce a stripe; its thickness depending on the selected wheel width.

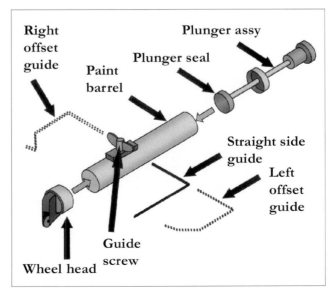

Drawing 10.29: The internal construction of the Beugler lining tool, also showing its guides

112 ■ CHAPTER 10: PAINTING

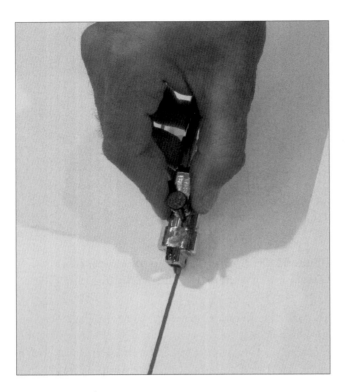

Photo 10.30 Hold the Beugler pen and smoothly draw it towards you. Use wrist action for curves

The lines tend to be slightly wider than the wheel and this is exaggerated when drawing curves, particularly small radius ones. Ensure the wheel itself is kept vertical unless producing a tapered line and move the wrist to paint any lines that are curved.

When lining ferrous metals, a magnetic guide strip can be fitted in place. For other materials, use masking tape as a guide but it should not form an edge to the line.

Bob Moore's Master Lining Pen, illustrated on Page 98, is easy to use and can give excellent results even in the hands of an inexperienced user. Fill the small paint reservoir and the pen is capable of producing around a metre or yard of line, depending on the line width. Use the guides to follow suitable edges.

An alternative is to use a draughtsman's pen filled with an appropriate paint to draw the line. Paint must be fed into the pen's reservoir and the line drawn by pulling the pen slowly, at a constant speed, towards the handle with the tip tilted back around 15° to ensure a regular flow of paint.

Using a bow pen, the paint should be fed into the area between the prongs of the bow and the bow adjusted to give a line of the desired width. Any paint on the outside of the bow should carefully be wiped off. Bow-pen compasses can be used for painting circular lines. In all cases rulers and home-cut templates will help to provide straight or curved lines in their correct places.

If the line is wider than the lining tool or pen will draw, mark each side of the line with the tool or pen and then fill between them with a fine lining brush. Any lining tool is very sensitive to paint viscosity and the head needs frequent flushing; after each line has been drawn.

Photo 10.31: A steam lorry with intricate yellow and black lining on the cab and mudguards

METAL FINISHING TECHNIQUES 113

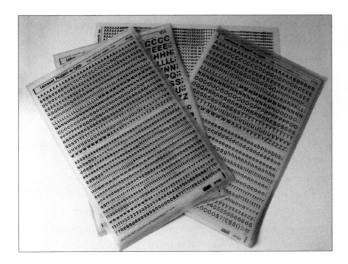

Photo 10.32: Letraset rub-on lettering comes in a wide range of sizes and fonts

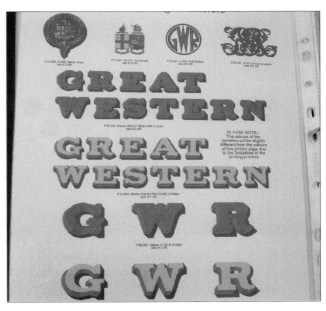

Photo 10.33: Some of the wide range of specialist transfers available from Precision Paints

Practise is the key to successful use. Corners are without doubt the hardest to line accurately. When using any lining device it is ideal if the surface is flat and horizontal. However, this is rarely the case with engineering models. In the case of a boiler, ideally it should be held in a frame that allows the model to be rotated as the line is drawn, with a wire wound around the boiler in the appropriate place to act as a guide.

Lining guides
Rulers for straight lines and cut-out Plasticard templates for curves are often needed. Then the Beugler or the Bob Moore adjustable lining guides can run along them.

LETTERING AND LOGOS
Few model engineers are able to sign write, as painting letters and intricately designed badges is a very skilled task. Fortunately, there are several practical alternatives.

One solution is to use 'press-on' lettering that was first popularised by Letraset and is now widely available in many different fonts, sizes and colours. They are easily fixed in place by removing the backing sheet, carefully locating them in position and then rubbing over the individual letter or logo with a ball-point pen.

Another option is to use water-slide transfers (decals) that are produced to suit a wide range of historic scale models. Careful removal of any clear film with a sharp scalpel may be necessary. Do follow any instructions that are provided. Soak water-slide transfers in warm water (usually around half a minute in the water) and use their backing to position them in their correct place. They are relatively fragile at this stage. First slide away the backing paper, removing any air bubbles, then excess water by gentle hand pressure on a piece of paper towel. Protect the transfers after application with a thin coat of varnish over just the transfer and its edges. These coatings do, however, tend to leave a tint on the base paint.

It is also quite feasible to use a computer to produce the art work and then print it on water-slide transfer paper. This comes with two different forms of backing; white and transparent. The former give far better coverage over the base paint of the model. Several companies offer lettering and logos that are produced to match the requirements of a wide range of scale engineering models. A less realistic solution is to use commercial lettering cut from self-adhesive plastic film or to produce the required lettering and logos on a PC and print them out on special white self-adhesive film. In this case, great

Photo 10.34: Lettering of different sizes and fonts on the side of a private-owner wagon

Photo 10.35: The majority of coats of arms are professional transfers or produced on a computer

care will be necessary to cut carefully around the edge of the print with a scalpel to avoid leaving an edge of plastic film that will ruin the effect.

For those determined to paint on the lettering themselves, there are sign writers' brushes available from artist's suppliers that have chisel ends. Then what is needed is careful outlining with a soft pencil of the lettering to be painted, the right brush and thickness of paint, a steady hand and plenty of practice. And a stick with a padded end is a useful adjunct to provide some wrist support.

CLEANING AND STORAGE

Brushes and spraying equipment must be completely cleaned with suitable thinners when changing the paint type or colour and immediately after use. And the thinners should be what was used to thin the paint; acetone for cellulose, white spirit or turpentine for enamels and polyurethanes. Two-part paints must always be cleaned off before the paint has a chance to harden. Once the chemical process has set these paints, the strength of the bond makes them extremely difficult to clean from spraying equipment and impossible to remove from brushes. Quality brushes and spraying equipment are expensive to buy but as long as they are thoroughly cleaned immediately after use, they will last a long time if not a lifetime.

When cleaning a brush straight after use, wipe any remaining paint from the brush with a rag, preferably lint-free, and then dip the brush in thinners to remove all the paint. It is astonishing how much colour will come out of the brush head. Make sure that the base of the head is clean. Then wash the brush with soap, not detergent, working up a good lather and rinsing the brush with warm but not hot water until there is no sign of colour. Ensure every last trace of soap has been removed, dry the brush and its handle and use fingers to shape it. Then stand it head up to allow the hair to dry. Never leave a brush resting on its bristles.

The build up of particles of pigment at the base of a brush is a common reason why it no longer forms a point. If allowed to accumulate, the paint pushes the hairs apart and prevents the point forming. Never allow paint to dry on any brush as then the only solution is to use paint stripper that is likely to ruin its shape.
Once well cleaned and dried (to avoid mildew), store natural animal-hair brushes in a container with a tight-fitting lid to avoid them becoming moth-eaten.

The most common problem that blocks spraying equipment is dried paint in the tip, nozzle or paint tube. It is worth spraying into an empty container with the paint supply turned off to remove the bulk of the paint from the innards. After turning off the compressor, point the spray at the container, release any residual pressure in the system and disconnect the hose. Then remove the reservoir, empty any remaining paint, thoroughly clean it, any lids or gaskets and the nozzle with thinners. Next, pour a small amount of clean thinners into the reservoir and refit it. Re-attached the hose, turn on the compressor and spray the thinners into an old canister until just clean solvent is being sprayed. Turn off the compressor again, disconnect the hose, empty any remaining thinners from the reservoir, wipe over the outside and store the equipment safely away.

Store paint in dry, cool and frost-free conditions. Most paints are highly inflammable so that they should be kept in a safe place. After only part of a container of paint has been used, it is often not easy to completely reseal the tin so the part-used paint can be stored for a lengthy period. Ensure that the lid is secure and turn the container upside down for a short time so that the remaining paint can seal any small air holes around the lid. Then store the tin the right way up. However, despite doing this, part-used paint that has been stored for more than a couple of weeks may well be unsuitable for re-use. It will at least require extensive re-mixing and straining and, particularly if a skin has formed, it may still prove to be unsatisfactory.

CONCLUSIONS

This book has attempted to examine the enormous number of different ways in which metals may be finished; both those parts that are visible and those hidden in places like cylinders and bearings. Some of the method apply to almost every model or full-size item likely to be worked on by a model engineer. Some involve very specialised techniques that will only rarely be needed.

Hopefully, the amount of information given in this book will assist beginners and experienced model engineers to understand the key benefits of the many different finishing techniques described. It may also encourage them to try alternative finishes during the construction of their projects.

Perhaps a careful study of the processes described in this book, allied to plenty of practice in the workshop, will help the average engineer to produce a higher standard of model. While a theoretical understanding can be useful, in almost every case it is the practical implementation that is the key to success.

The paint work of any new model is at risk during final assembly after completing the component parts. Possibly the time when the finish of a model is most likely to get damaged is when it is being moved from place to place, so always take great care. However, paint work is also easily spoiled if it is overheated or if it is given inadequate time to dry after application.

Most metal exposed to the atmosphere will rapidly tarnish and working models will inevitably accumulate all sorts of dirt. The task of keeping surfaces that have been polished or plated in tip-top conditions requires both care and forethought. Always use a clean duster when applying polish to avoid scratching the surface. The same is true when cleaning or polishing paint work. Internal combustion, hot-air and steam engines are best cleaned once they have cooled right down, removing any products of burning the fuel as well as any oil, water, dust and dirt.

The information in this book should enable builders to choose appropriate finishes for the various parts of

Photo C1: The job is done; the model painted, wheels lined, brass polished, bare metal protected and moving parts reamed, honed or lapped as neccessary. Steam is raised; she is ready to go

Photo C2: All the hard preparation has paid off as the driver makes some relaxing circuits of the track

their models to provide a scale appearance. It should also allow them to match an existing look whenever repairs are made. Finally, it may help them to protect the various different types of metal that are inevitable used.

Most of the techniques described can safely be employed by even the tyro model engineer but where potentially dangerous methods are suggested, warnings of the hazards have been included.

In many instances, particular surface speeds have been suggested for processes such as grinding and polishing. To aid calculation of the required rotational speed for any particular component size, a simple formula is given both in metric and imperial units and is included as Table C3. This relates revolutions per minute (RPM), surface speed and component diameter. The calculation can readily be carried out on a pocket calculator or even with just a pencil and paper.

$$RPM = \frac{\text{Required surface speed in metres/minute} \times 1000}{3.14 \times \text{diameter in millimetres}}$$

OR

$$RPM = \frac{\text{Required surface speed in feet/minute} \times 12}{3.14 \times \text{diameter in inches}}$$

Table C3: Formulae for calculating RPM for any particular diameter and required surface speed

USEFUL CONTACTS

3M Select Store, 3M Centre, Cain Rd, Bracknell, RG12 8HT, UK. www.3mselect.co.uk

Arc Euro Trade, 10 Archdale St, Syston, Leicester, LE7 1NA, UK. www.arceurotrade.co.uk

ATI Garryson Ltd, Spring Rd, Ibstock, Leicestershire LE67 6LR, UK. www.atigarryson.co.uk

Axminster Power Tool Centre Ltd, Unit 10, Weycroft Av, Axminster, Devon, EX13 5PH, UK. www.axminster.co.uk

Badger Air-Brush Co, 9128 W Belmont Av, Franklin Pk, IL 60131, USA www.badgerairbrush.com

S.B. Beugler Co, 3667 Tracy St, Los Angeles, CA 90039, USA. www.beugler.com

C & M Topline, Vibratory Finishing Equipment, 5945 Daley Street, Goleta, CA 93117, USA. candmtopline.com

Chronos Engineering Supplies, Unit 14, Dukeminster Estate, Church St, Dunstable, LU5 4HU, UK. www.chronos.ltd.uk

Chester Machine Tools Ltd, Clwyd Cl, Hawarden Ind Pk, Hawarden, Nr Chester, Flintshire, CH5 3PZ, UK. www.chesteruk.net

Combination, Unit 37, Peel Industrial Estate, Chamberhall St, Bury, Lancs, BL9 OLU, UK. www.combicolor.co.uk

Daler-Rowney, Daler-Rowney House, Peacock La, Bracknell, Berkshire, RG12 8SS, UK. www.daler-rowney.com

ITW DeVilbiss, Ringworld Rd, Bournemouth, Dorset, BH11 9LH, UK. www.itwifeuro.com

Dinitrol Rejel Automotive Ltd. Rejel Ho, Murdock Rd, Bedford, MK41 7PE, UK. www.dinitrol.com

Dushan Grujich horology web page: http://au.geocities.com/dushang2000

Frost Auto Restoration Techniques Ltd, Crawford St, Rochdale, Lancs OL16 5NU, UK. www.frost.co.uk

Hammerite Products, ICI Paints plc, Customer Care Centre, Wexham Rd, Slough, Berkshire, SL2 5DS, UK. www.hammerite.com/uk

Henkel Technical Services, Road 5, Winsford Industrial Estate, Winsford, Cheshire, CW7 3QY, UK. www.henkel-technical-services.co.uk

Home and Workshop Machinery, 144, Maidstone Rd, Foots Cray, Sidcup, Kent, DA14 5HS, UK. www.homeandworkshop.co.uk

USEFUL CONTACTS

Isopon U-Pol Products, Denington Industrial Estate, Denington Rd, Wellingborough, Northants, NN8 2QP, UK. www.u-pol.com

Letraset Ltd, Kingsnorth Industrial Estate, Wotton Rd, Ashford, Kent, TN23 6FL, UK. www.letraset.com

Liberon Ltd, Learoyd Rd, Mountfield Industrial Estate, New Romney, Kent, TN28 8XU, UK. www.liberon.co.uk

Machine Mart Ltd, 211 Lower Parliament St, Nottingham, NG1 1GN, UK. www.machinemart.co.uk

Metalblack, Delway Technical Services, 192 Seabank Rd, New Brighton, Wallasey, Wirral, L45 5AG, UK.

Micro-Surface Finishing, 1217 West Third St, PO Box 70, Wilton, IA 52778, USA. www.micro-surface.com

Morgan Clock Co, 815 Century Dr, Dubuque, IA 52002, USA. www.morganclock.com

Myford Ltd, Wilmot La, Chilwell Rd, Beeston, Nottingham, NG9 1ER, UK. www.myford.com

Phoenix Precision Paints Ltd, PO Box 8238, Chelmsford, Essex, CM1 7WY, UK. www.phoenix-paints.co.uk

Plasti Dip UK, Unit 1, Harvesting La, East Meon, Petersfield, Hampshire, GU32 1QR, UK. www.plastidip.co.uk

Plastic Padding Chemical Metal: Henkel Ltd, Hatfield (Consumer Adhesives), Apollo Ct, 2 Bishop Sq Business Pk, Hatfield, Herts AL10 9EY, UK www.loctite.com

Polly Model Engineering Ltd, Bridge Court, Bridge St, Long Eaton, Nottingham, NG10 4QQ, UK. www.pollymodelengineering.co.uk

POR-15 Inc, PO Box 1235, Morristown, NJ 07962, USA. www.por15.com

Reeves 2000, Appleby Hill, Austrey, North Warwickshire, CV9 3ER, UK. www.ajreeves.com

RustSolutions, The Garage Place, 24/26 Salisbury Rd, Gravesend, Kent, DA11 7DE, UK. www.jenolite.com

Sculpt Noveau. www.sculptnouveau.com

Shesto Ltd, Unit 2, Sapcote Trading Centre, 374 High Rd, Willesden, London, NW10 2DH, UK. www.shesto.co.uk

The Milliput Company, Unit 8, Marian Mawr Industrial Estate, Dolgellau, Gwynedd, LL40 1UU, UK. www.milliput.co.uk/

Warco, Warren Machine Tools, Warco House, Fisher La, Chiddingfold, Surrey, GU8 4TD, UK. www.warco.co.uk

WD-40 Co Ltd, PO Box 440, Kiln Farm, Milton Keynes, MK11 3LF, UK. www.wd40.co.uk

Windsor & Newton, Whitefriars Av, Harrow, Middlesex, HA3 5RH, UK. www.winsornewton.com

W&W Co Inc, No 104, Sec 1, Fen Liao Rd, Lin Kou, Taipei (24452), Taiwan, R.O.C. www.ww2.com

INDEX

A
Acrylic, 103, 104
Aluminium oxide, 12, 14, 20, 24, 28, 31, 36, 40-48, 51
Angle grinders, 21
Anodising, 90, 91
Antiquing fluid, 92
Applying patinas, 91

B
Badger, 95
Bakelite, 50
Belt sanders, 24
Bench grinders, 16, 17
Beugler, 97, 111, 112
Blemishes, 111
Bluing, 87
Bob Moore, 98, 112
Bonded abrasives, 12, 23, 27
Bonding, 13
Bright dips, 8
Broaching, 61-67
Brushed finish, 82
Brushes, 70, 82, 94, 95, 113
Brushing, 82, 102, 107
Buffing, 33, 34, 36, 37, 39, 40
Buffing aluminium, 39
Buffing brass, 39
Buffing castings, 40
Buffing machines, 34
Buffing steel, 40
Buffing zinc, 40
Burnishing, 69-72

C
Camouflage, 110
Carborundum, 12, 24, 82
Castings, 26, 31, 39, 40, 41, 43, 61, 90, 100, 106
CBN, 12, 14, 23, 26, 57
Cellulose, 10, 94, 95, 98, 102
Chasing, 78

Chemical Metal, 100
Cleaners, 7
Cleaning, 7, 51, 110, 113
Cloth, 24, 97, 99
Coated products, 24
Cold bluing, 88
Colour, 33-37, 103, 104, 107, 108, 110, 111, 113
Compounds, 36, 37
Compressors, 96, 97
Contamination, 110
Convolute wheels, 28
Crocus, 12, 36, 51, 70
Cromium, 89
Cutting tools, 9

D
Dalon, 94
De-burring, 6
DeVilbiss, 95
Diamond, 12, 14, 26, 43, 44, 50, 51
Disc sanders, 24
Dressers, 15, 19
Drilling, 6

E
Electro-polishing, 41, 42
Emery, 7, 12, 23, 25, 27, 36, 44, 48, 51, 70, 94, 110, 111
Enamel paints, 102
Engine turning, 81, 82
Engraving, 79-81
Epoxy paints, 103
Equalising lapping, 48
Etching, 86, 101
Expandable machine reamers, 58
Exposure, 9
External lap, 49, 53

F
Filing, 6, 7
Fillers, 100, 101

Flap wheels, 25, 26, 28
Flat sheets, 24
Flat surfaces, 20, 27, 29, 45, 49, 52, 65, 70, 73-75, 104
Fly cutting, 6
Form lapping, 53
Frisk film, 97, 108
Frosting, 77, 85

G
Garnet, 24, 31
Garryflex blocks, 27
Gilding, 92
Gold, 1, 105
Grains, 12, 13
Grinding, 11-21, 33
Grinding machines, 16

H
Haematite, 89
Hand burnishing, 70, 71
Hand reamers, 57, 59
High-temperature paints, 105
Holtzapffel, 81
Hones, 43-48
Honing, 43-48
Hot bluing, 87
HSS, 14, 17, 60, 84
Humbrol, 102
HVLP, 96

I
Internal broaches, 64, 65
Internal lap, 50
Isopon, 100

J
Jacot drum, 72
Jade oil, 89, 92
Jax, 92
Jigs, 17, 45

K
Knurling, 83, 84
Koolblak, 88

L
Lacquer, 104
Lapping, 43, 48-53
Lapping compounds, 50, 51
Lapping powder, 51
Lead, 41, 50, 91, 101
Lettering, 112, 113
Liberon, 89, 92

Lining, 97, 98, 111, 112
Logos, 112, 113

M
Machine reamers, 57, 59
Masking, 107, 108
Mesh, 12, 13
Metalblak, 88
Metallic paint, 105
Micro balls, 30
Micro-Mesh, 41
Micron, 41, 51, 73, 91
Micron-graded abrasives, 41
Mild abrasives, 36
Milling, 5, 73
Milliput, 100
Mini drill, 20
Mixing, 107
Morgan pivot polisher, 72
Morse taper, 56, 57, 59

N
Nitromors, 106
Non-woven products, 27, 28
Nylon, 23, 27, 28, 39, 50, 108

O
Oil stones, 44
Organic, 91, 94
O-rings, 53

P
Paint choices, 101
Paint stripper, 10, 79, 88, 94, 106, 114
Peening, 77
Pens, 98, 112
Perma-Grit, 27, 28
Phosphor bronze, 70
Phosphoric acid, 7, 8, 10, 41, 82
Pickles, 8
Planishing, 77
Plasti Dip, 106
Plastic coating, 106
Polishing, 33, 34
Polyurethane, 103, 104
Primer, 100
Primer fillers, 100

R
Reaming, 55-60
Repoussé, 78, 79
Rouge, 12, 36, 37
Runs, 110

Rust, 7, 99, 106
Rust removal, 7
Rust-preventing paint, 106
Rusty metal, 99

S
Sagging, 110
Sand blasting, 30, 31
Sanding, 23-31
Sawing, 6
Scraping, 73-75
Sculpt Nouveau, 92
Shape and size, 15
Shellac, 104, 105
Silicon, 90
Silicon carbide, 12, 14, 17, 24, 27, 28, 31, 40, 41, 44, 51
Solvent, 10, 85, 87, 94, 95, 98, 100-107, 110, 114
Spray guns, 96
Spraying, 10, 95, 97, 108
Spraying booth, 97
Spraying equipment, 95
Stainless steel, 7, 8, 40, 43, 70
Steel wool, 29
Stoving paints, 105
Super abrasives, 23, 26
Surface broaches, 64, 65
Surface grinders, 16, 20
Surface preparation, 98

T
Tack cloth, 97, 99
Taper reamers, 59
Tectrate steel-blackening salts, 88

Teflon, 95
Thinners, 10, 94, 97, 98, 100-102, 105, 106, 108, 109, 113, 114
Tool and cutter grinders, 17
Tool-post grinders, 19
Tripoli, 12, 37
Truing and dressing wheels, 15
Tumble burnishing, 72
Tungsten carbide, 17, 23, 26, 27
Two-part, 103, 104, 113
Types of broach, 62

U
Unified wheels, 28, 29
Using a scraper, 74
Using broaches, 65, 66
Using reamers, 59, 60

V
Varnish, 4, 10, 40, 87, 93, 94, 98, 99, 103-106, 112
Vaseline, 51

W
Water stones, 44
WD40, 47, 82
What is being modelled, 9
Wheels, 15, 35
Whetstones, 21
Wire wool, 29
Working surfaces, 9

Z
Zinc, 40, 101